Surface Mining Machines

Eugeniusz Rusiński · Jerzy Czmochowski
Przemysław Moczko · Damian Pietrusiak

Surface Mining Machines

Problems of Maintenance and Modernization

Springer

Eugeniusz Rusiński
Faculty of Mechanical Engineering—Department of Machine Design and Research
Wroclaw University of Science and Technology
Wrocław
Poland

Przemysław Moczko
Faculty of Mechanical Engineering—Department of Machine Design and Research
Wroclaw University of Science and Technology
Wrocław
Poland

Jerzy Czmochowski
Faculty of Mechanical Engineering—Department of Machine Design and Research
Wroclaw University of Science and Technology
Wrocław
Poland

Damian Pietrusiak
Faculty of Mechanical Engineering—Department of Machine Design and Research
Wroclaw University of Science and Technology
Wrocław
Poland

ISBN 978-3-319-47790-9 ISBN 978-3-319-47792-3 (eBook)
DOI 10.1007/978-3-319-47792-3

Library of Congress Control Number: 2016956182

Printed on acid-free paper

This Springer imprint is published by Springer Nature
The registered company is Springer International Publishing AG
The registered company address is: Gewerbestrasse 11, 6330 Cham, Switzerland

Preface

Heavy machines are the main equipment in surface mines around the world. With the development of new technologies, surface mining has also been evolving. Among the leading drivers for change and development in this area are the increasing requirements to improve safety, to reduce the environmental impact and to reduce operating costs. The current and future challenge involves modernizing and maintaining old, worn-out machines and designing new structures.

The construction and operation costs of these machines are counted in hundreds and even millions of euros. These machines are designed for operation 24/7, so when they are out of operation due to failure, they generate huge losses. Occasionally, catastrophic failures happen, which can have negative economic consequences, but which, nevertheless, cannot be compared to the damage to human health and life.

In addition to rock processing, mega machines are mainly used in lignite surface mining. Despite the global trend to obtain energy from renewable sources, a significant percentage of energy production will still come from conventional sources of fuels for many decades. This is dictated by the need to ensure a cheap and stable source of energy. In many countries, the amount of energy consumed from brown coal is significant (e.g., Germany, Poland, Serbia, Greece). In emerging markets (Brazil, India, China), the lignite mining industry is rapidly developing. The most common problem, however, is the age and thus overuse of mining and processing machines (European market). Many of them are over 35 years old, and hence their operational life is close to wearing out. This is a considerable challenge for the users. However, users have gained unique experience over the years, which is an excellent base for developing further operational guidelines and an invaluable source of knowledge that can be used in designing new structures. The research and methods presented in this book have been developed and reviewed in cooperation with the users of such machines. Therefore, the presented content is not a collection of theoretical considerations, but rather of practical guidelines that can be applied in reality. The developed solutions use both the already known research tools and the most modern scientific methods. What is most important, however, is that they are practical and modern.

The examples of applications and methods presented in this book have contributed to the increase in the competitiveness and innovation capacity of the mining and rock processing industry. They are an important contribution to the development of knowledge of the construction and operation of heavy machinery. Above all, however, they have a wide practical application, which provides a more economical, efficient and safe operation.

Wrocław, Poland

Eugeniusz Rusiński
Jerzy Czmochowski
Przemysław Moczko
Damian Pietrusiak

Contents

About the Authors

Eugeniusz Rusiński graduated in 1975 with a Master of Engineering Degree in Mechanics and Machine Design from the Mechanical Faculty of Wrocław University of Technology. In the same year, he was hired as Assistant Lecturer at his *alma mater* and started his Ph.D. program, which he completed in 1980 and was awarded the Doctor of Science degree. In 1990, the Board of the Mechanical Faculty of Wrocław University of Technology awarded him the degree of Habilitated Doctor of Technical Sciences in machine construction and operation, specializing in theory of machines and mechanisms, for his dissertation entitled "*Microcomputer analysis of frames and superstructures of vehicles and construction machinery*". In 1993, he was promoted to the position of Associate Professor, and in 2001 to the position of Full Professor. In 1996, the President of Poland awarded him the title of Professor of Technical Sciences. He developed experience in Poland and abroad including, among others, in Germany, USA, India, Austria and Cuba.

In 1994, he was elected head of the Department of Computer-Aided Design (CAD), which, under his leadership, was transformed in 2013 into the Department of Machine Design and Research of Wrocław University of Technology. From 1995 to 1998, he was Deputy Director, and then from 2000 to 2005, Director of the Institute of Machine Design and Operation. In the years 2005–2012, he was Dean of the Mechanical Faculty. In the years 2008–2016, he was Vice-Rector for Research and Cooperation with Industry at Wrocław University of Technology.

He supervised nineteen doctoral dissertations. He is author of numerous scientific publications. He has received the Team Award of the Prime Minister of the Republic of Poland five times.

Jerzy Czmochowski graduated in 1978 with a Master's Degree in Engineering from the Faculty of Mechanical Engineering of Wrocław University of Technology. In 1985, he defended his doctoral dissertation entitled: "*Stability analysis of non-simply-connected three-layer plates using the finite element method*", under the supervision of Prof. Franciszek Romanow. After completing his doctoral studies in 1982, he started work as Senior Assistant Lecturer, and from 1989 to 2009 he was employed as research and teaching Assistant Professor at the Institute of Machine Design and Operation at Wrocław University of Technology (WrUT). In 2008, he was awarded the degree of habilitated doctor by the Mechanical Faculty of Wrocław University of Technology. On November 12, 2009, he was appointed to the position of Professor of WrUT. He supervised two doctoral dissertations. His achievements include papers in the fields of fracture mechanics, diagnostics of emergency states, dynamics of surface mining machines, thermal and mechanical analysis of power industry equipment and strength analysis of load-carrying structures of machines. He received the Award of the Prime Minister of the Republic of Poland for outstanding scientific achievements in 1999 and in 2011. In 2010, he was awarded the Gold Cross of Merit.

Przemysław Moczko graduated in Mechanical Engineering from Wrocław University of Technology, Faculty of Mechanical Engineering. In October 1999, he started the Ph.D. program in Design and Operation of Machines. The focus of his doctoral dissertation, which he defended in 2003, was on issues related to predicting the durability of elements of load-carrying structures. In 2015, he received the degree of habilitated doctor for his achievements in applying numerical and experimental methods in condition assessment and prediction of the durability of load-carrying structures of machines and in monitoring the processes and condition of low-speed machines. He is also co-author of significant technical achievements: first, the excavating unit cutting elements of surface mining machines; and second, the innovative fab wheel of the WPK-5.35 centrifugal fan. His scientific achievements are confirmed by numerous publications in national and international journals and by patents and pending patent applications. He also has the status of an international independent expert on assessment of the designs of specialist mining equipment, their implementation and compliance with required standards.

Damian Pietrusiak graduated in Mechanical Engineering from Wrocław University of Technology, Faculty of Mechanical Engineering. In October 2009, he started the Ph.D. program in Design and Operation of Machines. In 2013, he defended his doctoral dissertation entitled: "*Assessment of the dynamics of load-carrying structures of bucket-wheel excavators with the application of modal analysis*". Currently, he is Assistant Professor at the Department of Machine Design and Research. In his scientific work, he focuses on structural dynamics and statics of load-carrying structures as well as on the application of innovative engineering solutions in the heavy industry, the automotive industry and other sectors. He specializes in vibration and modal analysis with the use of numerical and experimental methods. He cooperates with different industries as contractor and manager of numerous research projects. He is an author of many research papers and patents/patent applications.

Chapter 1
Introduction

One of the cheapest methods of generating energy by burning lignite. This type of coal is mined in open-pit mines, first by excavating a layer of overburden and then the coal itself. Once the layer of coal deposit is uncovered, the overburden and lignite are excavated simultaneously by special machines. The process of excavating overburden uses bucket wheel excavators (Figs. 1.1, 1.2 and 1.3), mobile conveyors (Fig. 1.4), conveyors (Fig. 1.5) and spreaders (Fig. 1.6). The same machines are used for excavating lignite, but the coal is transported by belt conveyors to a stockpile at the power plant, where stacker-reclaimers are used (Fig. 1.7). The deepest layers of coal are excavated by bucket-chain excavators (Figs. 1.8 and 1.9). The advantage of these machines is that they can excavate above, below or on the working level. Bucket-chain excavators use a continuous pattern of buckets which move longitudinally and perpendicularly to the axis of the cutting bucket. In that case the main working movement of the buckets attached to the chain is a rectilinear movement, whereas the lateral movement is caused by the motion of the entire rig, which can be rectilinear or rotating. In bucket-wheel excavators the working mechanism consists of a bucket wheel with buckets arranged around its circumference. The working movement of the excavating mechanism is the rotation of the wheel, and the lateral movement is usually caused by the rotation of the bucket wheel boom. The cuts made by subsequent buckets mounted on the wheel have a characteristic crescent shape (Fig. 1.10).

The overburden or coal stripped by excavators is transferred onto belt conveyors mounted on the excavator, and then transported onto the belt conveyor located on the bridge (Fig. 1.2) or on the discharge boom attached to the excavator (Fig. 1.3) or on an additional mobile conveyor (Fig. 1.4). Next, the overburden is transported on overland belt conveyors and is initially stockpiled on an outside dump, and after the coal has been excavated, the overburden stockpiled at the mine site (Fig. 1.11).

While digging, multi-bucket excavators are subject to substantial static and dynamic loads. Dynamic loads result from the excavation process in which the buckets cut the ground generating cycling excavation force—especially high loads occur in the case of hard-mineable materials, rocks or dirt bands. Additional

E. Rusiński et al., *Surface Mining Machines*,
DOI 10.1007/978-3-319-47792-3_1

Fig. 1.1 SchRs 4600.30 bucket wheel excavator (Bełchatów Coal Mine)

Fig. 1.2 Bridge with a belt conveyor attached to a bucket-wheel excavator

dynamic loads include massive inertia forces which arise when the machine travels, rotates or changes the position of the bucket wheel boom.

Since surface mining machines are extremely large, complex and thus expensive, they should be able to operate for at least 30 years and often even up to 50 years.

Fig. 1.3 Bucket-wheel excavator for hard-mineable materials with a discharge boom

Fig. 1.4 PGOT-5500 mobile conveyor (Bełchatów Coal Mine)

Machine assemblies such as electric motors, gearboxes, mechanical or hydraulic clutches, ball or sliding bearings and ball joints are subject to natural wear, e.g. caused by friction between elements. Over time, these assemblies can also become obsolete as new solutions are introduced, including machines with better performance and assemblies that are lighter and cheaper to use. This is also the reason why such elements are replaced during modernization, which lowers the costs of operation, and, owing to technological advancement, offers more reliable machine assemblies. An example of such practices is the replacement of heavy spur gears in drive trains of the bucket wheel with much lighter planetary gears, which substantially decreased the load to the load-carrying structure.

Fig. 1.5 Overland belt conveyors on a lignite bed (Turów Coal Mine)

Fig. 1.6 Spreader (Turów Coal Mine)

Elements of these machines can fail (Fig. 1.12) to a lesser or greater due to: design faults, fatigue of material (Fig. 1.13), technological faults, operating misuse or use of these machines in inappropriate conditions. Design faults usually consist in notches, i.e. areas of stress concentration caused, for instance, by an insufficient cut radius in the drive shafts in places where there is a sudden (step) change in stiffness. Imperfections in the geometric shape can also be due to the applied calculation methods and the adopted calculation model. The calculation methods that were used to design machines which are now several decades old, were not as accurate as today's methods. Moreover, the current standards are based on many years of experience in designing such machines and take into account many more factors that are necessary to properly dimension such elements.

Fig. 1.7 ŁZKS 1600 Stacker-Reclaimer (Bełchatów Coal Mine)

Fig. 1.8 Rs-560 bucket-chain excavator (Turów Coal Mine)

The load-carrying structure is one of the elements which are never replaced in the entire lifetime of a machine. The load-carrying structure is expected to have almost infinite durability. However, due to the imperfect calculation methods and design imperfections, smaller or larger fractures do occur. Moreover, environment stress causes gradual degradation in the form of corrosion which weakens the load-carrying elements.

Fig. 1.9 ERs 710 bucket-chain excavator (Bełchatów Coal Mine)

Fig. 1.10 Cuts in overburden made by a bucket-wheel excavator

In order to extend the durability of load-carrying structures, machines that operate for long periods of time are thoroughly analyzed. Testing of the load-carrying structure includes detection of material discontinuity, fractures e.g. using non-destructive methods (NDT). Such testing should be performed periodically in areas that are particularly prone to fatigue cracks. Today's modern calculation methods based on accurate models of the systems under investigation can fairly accurately identify places of stress concentration, i.e. places particularly prone

Fig. 1.11 Overburden stockpile at mine site

Fig. 1.12 Fatigue crack of a half-axle in the crawler drive mechanism of a spreader

to fracture origination. By using shell or solid elements for calculation for the finite element method (FEM) it is possible to determine local stress distributions, especially in places where load-bearing beams are connected with pylons or bracings.

When evaluating the state of stress it is important to identify the actual loads acting on the load-carrying structure. In the case of surface mining machines these loads are defined by the DIN 22261 standard in the form of overload coefficients. In practice these values are exceeded, which means that the actual loads are often

Fig. 1.13 Collapse of the discharge boom onto the ZGOT—mobile transfer conveyor, due to fatigue fracture of the support pin

higher than those given in the standard. Therefore tests of machines after long-term operation should comprise measurements of vibrations of the entire machine and of the loads acting on the excavating unit. Dynamic forces identified using this method are included in computer simulations and thus offer better stress assessment.

Based on these analyses and tests of the machine, a prediction is made regarding its further operation. Structural changes are recommended to reinforce places characterized by high stress levels, or places in which the onset of a fracture process was identified. Recommendations are made regarding areas which should be periodically monitored. The purpose of such activities is to extend the durability.

This book describes the problems related to the preservation of the technical condition of surface mining machines. The second chapter presents the main problems in preserving the technical condition of load-carrying structures of bodies, undercarriages and elements of such machines. The subsequent chapter focuses on methods of evaluating the technical condition, based on experimental and numerical methods, which are illustrated with examples of analyses performed for bucket-wheel excavator systems. This is followed by an analysis of fatigue and evaluation of durability of selected surface mining machines. The last chapter presents selected examples of modernization methods for machines which were in operation for a long period of time.

Chapter 2
Main Problems Related to the Operation and Maintenance of Mega Machines

2.1 Introduction

There is a natural tendency to extend the operating time of machines, especially the complex and expensive ones, in order to minimize the costs of new investments and to avoid the downtime that is required to construct new machines. In the case of surface mining machines, this problem is much more tangible because giant earthmovers are the largest mobile machines built by man, which affects the cost and time needed to design and construct them. Moreover, in most mines, excavators are scheduled to operate until the mine is exhausted, which, in most cases, results in a natural need to extend the service life of the existing machine fleet.

Over time, this state of affairs increases the risk of failure in components and assemblies, which often causes long-term downtime for repairs or even dangerous accidents and disasters [1, 2], which, in turn, can result in the machine being removed from service or scrapped. Such events may be prevented by performing periodic inspections and tests on components of mega machines, especially in those subassemblies whose damage is dangerous for the entire machine [3]. The conducted research not only leads to useful conclusions related to their operation but also helps to expand knowledge on surface mining machines. By thoroughly analyzing new problems and issues related to the operation of surface mining machines, scientists, in collaboration with machine users, look for solutions to these new questions and thus expand the frontiers of knowledge. This primarily increases the safety of operation and extends the service life of mega machines.

E. Rusiński et al., *Surface Mining Machines*,
DOI 10.1007/978-3-319-47792-3_2

2.2 The Design of Surface Mining Machines

Safe operation and condition maintenance of surface mining machines requires excellent knowledge of their structure in terms of load-carrying structures and mechanical components, as well as of the control and power systems. The design of such machines is determined by the functions that are performed during the excavation of minerals, their transport and use in the production process, and their stacking or storage [4]. The following section covers the basic information on the structure of bucket-wheel and bucket-chain excavators and spreaders.

2.2.1 Bucket-Wheel Excavators

The general classification of bucket-wheel excavators is related to their size. According to this criterion, there are three basic groups of machines:

(a) Compact bucket-wheel excavators—these bucket-wheel excavators have a theoretical capacity of up to 5000 m^3/h and are the smallest, though in recent years there has been a tendency to construct larger machines of this type. They are characterized by a compact load-carrying structure and are equipped with hydraulic hoisting systems for the excavating and discharge booms. They are not equipped with a classic counterweight boom but the counterweight is located on the superstructure platform. The compact construction of such machines makes them suitable for operation in those areas of mines that are not easily accessible or in conditions where excavating is difficult due to the cutting resistance forces. They usually have a two-crawler undercarriage, which makes them easier to maneuver on the one hand, but limits their global stability on the other. Figures 2.1. and 2.2. show examples of different sizes of compact bucket-wheel excavators.

Fig. 2.1 KWK205L compact bucket-wheel excavator—designed by SKW Zgorzelec

Fig. 2.2 SRs(H) 1050.23/2.0 compact bucket-wheel excavator—designed by TAKRAF TENOVA (TAKRAF marketing materials)

(b) C-frame bucket-wheel excavators—mid-size bucket-wheel excavators with a theoretical capacity of up to 10,000 m^3/h (SchRs 4600.30), though the average theoretical capacity of such machines is approximately 5000 m^3/h. Their characteristic structural feature is the shape of the platform, pylons and the counterweight boom, which form the so-called C-frame. In contrast to compact excavators, these machines have a counterweight boom with a ballast permanently fixed to its end. The bucket-wheel boom is hoisted by means of a cable system with the main winch located on the counterweight boom. Some machines of this type (e.g., SRs 1200, SRs 2000), have an additional mast for the bucket-wheel boom, which is connected to the boom's head by means of a rigid tie rod. The cable-based boom hoisting system is located between the winch and the mast head. These machines are defined as mid-sized, which makes them suitable for operation in those areas of mines that are moderately difficult to reach. They can also operate in difficult excavating conditions in terms of cutting resistance forces but this requires a special design of their load-carrying structure and drive systems. One example is the KWK910L excavator designed by SKW Zgorzelec. The undercarriages of these machines are usually composed of three crawler tracks, or sometimes even four (SRs 1200), which makes them difficult to maneuver on the one hand, but provides good global stability on the other. Figures 2.3, 2.4, 2.5, 2.6 and 2.7 show examples of different sizes of C-frame bucket-wheel excavators.

(c) Large bucket-wheel excavators—these are the largest mobile bucket-wheel excavators with a theoretical capacity of up to 16,000 m^3/h. Their characteristic structural feature is that they have two main superstructure boom units: the excavating boom and mast interconnected by means of steel cables, and the counterweight boom unit, also with a mast, interconnected with similar steel cables. The two mentioned units are attached to the superstructure's slewing platform supported by a slewing bearing, which rests on an

Fig. 2.3 C-frame bucket-wheel excavator—BWE700L

Fig. 2.4 C-frame bucket-wheel excavator—BWE1400L

Fig. 2.5 C-frame bucket-wheel excavator—SRs 1200

Fig. 2.6 C-frame bucket-wheel excavator—SRs 2000

Fig. 2.7 C-frame bucket-wheel excavator—KWK910L

undercarriage portal frame and other elements of the tracked undercarriage. Moreover, these machines often have a ballast that can move along the counterweight boom depending on the positioning of the excavating boom. The tracked undercarriages have a three-point support design. However, the crawlers in individual supports are often multiplied in order to obtain the appropriate pressure of crawler pads on the ground. This is due to the fact that the mass of such excavators may exceed 10,000 tons. Figures 2.8 and 2.9 show examples of large bucket-wheel excavators.

2.2.2 *Bucket-Chain Excavators*

Multibucket chain excavators are used relatively rarely in operation due to their complicated structure and low excavating capacity. Nevertheless, they have one fundamental advantage over bucket-wheel excavators, namely a high range of excavation below the bench level in relation to the size of the machine. This is why they can be used in complex mining sites and in places where the pit face ends without the need to excessively enlarge the pit. These excavators are built in a similar fashion to bucket-wheel excavators in terms of their tracked undercarriage and counterweight and discharge booms. They are usually equipped with two bearings that control the slewing of the superstructure. Smaller machines have a

Fig. 2.8 Large bucket-wheel excavator—SchRs 4600

Fig. 2.9 Large bucket-wheel excavator—SchRs 4000

construction similar to that of compact bucket-wheel excavators. In such cases, a superstructure moment slewing bearing is used, whereas the counterweight is located on the superstructure platform. All bucket chain excavators are equipped

Fig. 2.10 Multibucket chain excavator—ERs710

with a chain composed of buckets connected by chain links. Such chain is mounted on a boom that consists of several segments, each with an independent hoisting cable system with winches mounted to the counterweight boom or to the superstructure platform. Such a construction makes it possible to excavate with the entire length of the chain boom, whose shape can be adjusted to the requirements of the digging technology. Examples of bucket chain excavators are shown in Figs. 2.10 and 2.11.

2.2.3 Spreaders

Conveyor spreaders are used in surface mines to pile the extracted overburden, which is collected on external or internal stockpiles at the surface mining site. External stockpiles are separated areas, usually near the surface mine, where overburden is stacked in stockpiles. Internal stockpiles are those parts of the mining site where extraction has been completed. Materials that are primarily stockpiled include overburden extracted by excavators or by-products of coal combustion in power plants, i.e. ash, which is usually transported from energy blocks by means of conveyor belts or, less frequently, by railway or trucks.

Fig. 2.11 Multibucket chain excavator—Rs560

Fig. 2.12 Spreader—A2RsB12500

Spreaders are similar in structure to excavators, but due to their function they are not equipped with an excavating unit. Instead, they have a discharge boom that piles the dredged material on stockpiles. The capacity of spreaders reaches up to 20,000 m^3/h of overburden. Examples of conveyor spreaders are shown in Figs. 2.12 and 2.13.

Fig. 2.13 Spreader—20000TPH

2.3 Main Mechanical and Structural Subassemblies of Surface Mining Machines and the Problems that Occur During Their Operation

The operational problems of surface mining machines can be grouped according to the functions carried out in individual stages of the mining process, in the case of excavators, and the spreading process, in the case of spreaders. The following section describes the main mechanical and structural subassemblies of bucket-wheel excavators and spreaders, in which operational problems occur. The main focus is on those structural and mechanical components that are subject to degradation processes and determine the machine's uptime and operational safety.

Excavating unit
This consists of the following elements, which are prone to operational problems:

- Bucket-wheel drive gear—this is subject to damage that is typical for spur gears, i.e. damage to the teeth, bearings, axles and shafts. Damage also occurs to the housing of gears due to their large size in this type of machines. Additionally, fractures can occur in moment beams whose function is to transfer the moment resulting from the cutting force to the load-carrying structure of the bucket-wheel boom. The problem of gear degradation can be improved significantly by using properly operating torque limiting couplings [5] (excluding hydraulic couplings, as these are not torque limiting couplings for impact loads), provided that correct and invariable coupling settings are maintained, which is also very important in the case of double or triple gear drives. Examples of damage to bucket-wheel drive gears are shown in Fig. 2.14.

- Bucket wheel shaft or axle—damage to this element is relatively rare. It is mostly prone to ultimate failure resulting from extreme overloads or to damage during disassembly in areas where loads are transferred, e.g., press fit between the drive gear and the shaft. Examples of damage to bucket-wheel shafts are shown in Fig. 2.15.
- Bucket wheel—this is mostly subject to fatigue damage, mainly around the wheel hub (attachment of the wheel to the shaft or axle), and at points where buckets are attached. Damage to this element of the excavating unit are relatively frequent. Examples of damage to bucket wheels are shown in Fig. 2.16.
- Cutting elements: buckets—these are mostly subject to the process of natural abrasive wear resulting from the excavation process [6]. In addition, there are

(a) (b)

(c) (d)

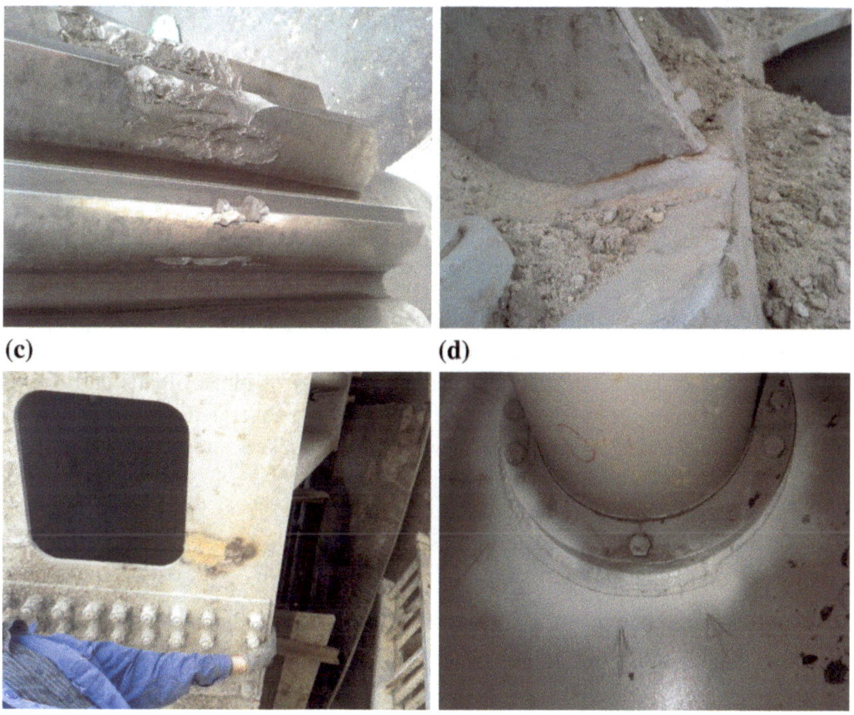

Fig. 2.14 Damaged elements of the bucket wheel drive gear in surface mining machines: **a** damage to spur gears of the drive gear in SRs 2000 excavators; **b** fractures on the housing of the gearbox in the SchRs 4000 excavator; **c**, **d** damage to the moment beam and its mount in the bucket-wheel drive gear on the SRs 1800 excavator

Fig. 2.15 Damage to the bucket-wheel drive shaft in the KWK 1200M excavator

(a) (b)

Fig. 2.16 Damaged areas of bucket wheels in excavators: **a** SchRs 4600 and **b** KWK 1200M

numerous cases of ultimate failure caused by dynamic overloads generated when a bucket strikes non-mineable material. Cases involving fatigue damage are rather rare since buckets are designed mainly for impact loads and operate for a short time before they are replaced and regenerated. Examples of the typical processes of wear and ultimate failure to buckets are shown in Fig. 2.17.

Superstructure slewing system

The superstructure slewing system consists of the following elements, which are subject to operational problems:

- Superstructure slewing drive gear—this is subject to damage typical for spur gears, which is further enhanced by the instability of loads generated during operation. This problem mainly pertains to excavators, whereas in the case of spreaders such a phenomenon is practically non-existent. In the case of excavators, the main load acting on the superstructure slewing gear drive is the lateral force generated during excavation. This is a time-varying load that is random in character and involves dynamic events that occur when the

Fig. 2.17 Wear and damage process of buckets

excavation unit is overloaded. This happens when a bucket strikes non-mineable material. Such drives are also subject to failures of their torque limiting couplings, whose purpose is to limit the force transmitted to the drive gear. When the machine operates under difficult and variable excavating conditions, the number of switching cycles of torque limiters can be sufficiently large to accelerate its degradation. Another practical problem involves ensuring proper adjustment of settings depending on the coupling type (friction, powder etc.) and the invariability of these values over time. There have been numerous cases involving problems in ensuring the same settings for individual or multiple drives (usually two or four superstructure slewing drives).

- Superstructure slewing ring—superstructure rotation drive is transmitted from the drive gear through a pinion onto the superstructure slewing ring. The problems that occur are typical of spur gears, especially in the case of excavators, which operate in more severe conditions than spreaders. An additional problem in the case of excavators can be damage to the elements that connect the slewing ring with the portal. These are usually connected by means of bolted connections, which are subject to loosening and, occasionally, cracking. This results from the heavy loads generated when overload occurs and the bucket-wheel drive is switched off, which in turn causes momentary overloads to the superstructure rotation drive system.

- The superstructure slewing bearing and slewing wheel support—the rotation of surface mining machines is carried out primarily through large slewing bearings or, less commonly, especially in older machines, through a slewing wheel support with a system of bogies, a raceway and bogie wheels [7]. In the case of slewing bearings, the main and numerous problems are associated with accelerated wear of the rolling elements and the raceway. In the case of the so-called hard raceways, wear mainly occurs as a result of pitting (less frequently, fretting) and the rapid degradation that follows, whereas in the case of the so-called soft raceways, wear is the result of plastic deformations and the "settling" of the bearing. Examples of damage to superstructure slewing bearings are shown in Fig. 2.18.
- Damage to superstructure slewing wheel supports of surface mining machines occur in the bearing joints of individual bogies and are caused by an uneven distribution of force on the bogie wheels. Other damage that typically occurs involves pitting of the raceway and wheels. These types of structural joints also generate damage to their supporting structures resulting from the lack of alignment in the transfer of loads between the bogie wheel and the raceway. As a result, the raceway and the top flange of the supporting structure experiences

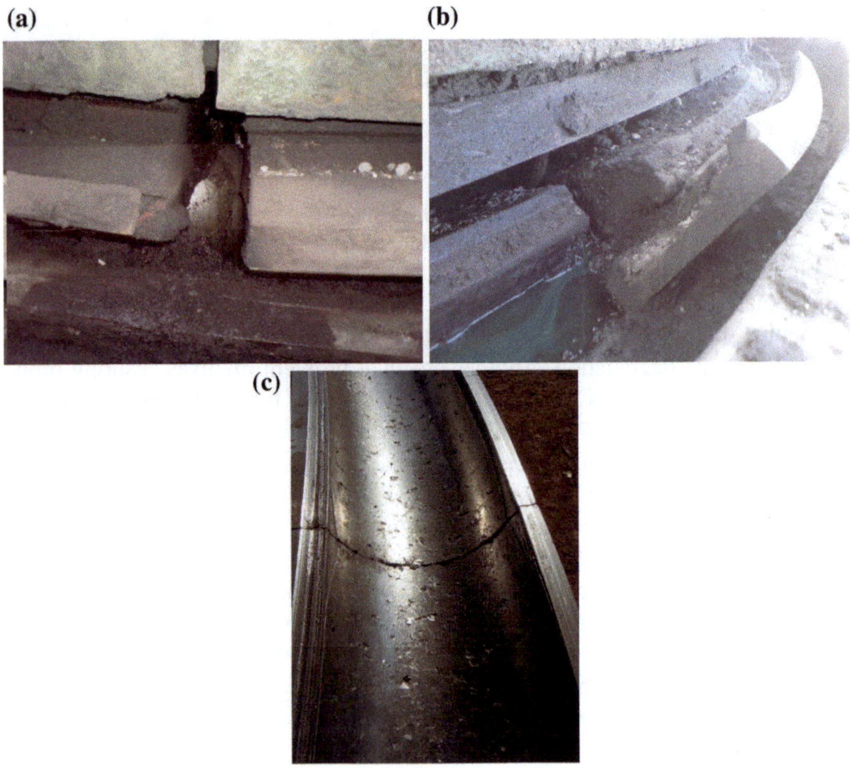

Fig. 2.18 Damage to the superstructure slewing bearing of surface mining machines: **a**, **b** plastic deformation of the bottom raceway; **c** pitting

local bending, which in consequence may lead to fatigue fractures. Examples of such phenomena are depicted in Fig. 2.19.

Undercarriage drive system

The undercarriage drive system consists of the following elements, which are subject to operational problems:

- Drive gear—this is subject to wear processes typical of spur gears, which, due to the intermittent character of operation, are not very intensive. Other problems include those related to the transmission of power to the drive wheel of the crawler track. The drive shaft transmits high torque and is subject to overloads resulting from the uneven surface on which the machine travels. Such a situation can cause damage to the shaft or to the connection with the output shaft of the drive gear. Another common problem in the operation of drive systems is the wear of drive wheels in the chain of the crawler track in the area where torque is transferred to the pad, as well as damage to this area in the form of fractures. Examples of damage to this system are shown in Fig. 2.20.

(a) **(b)**

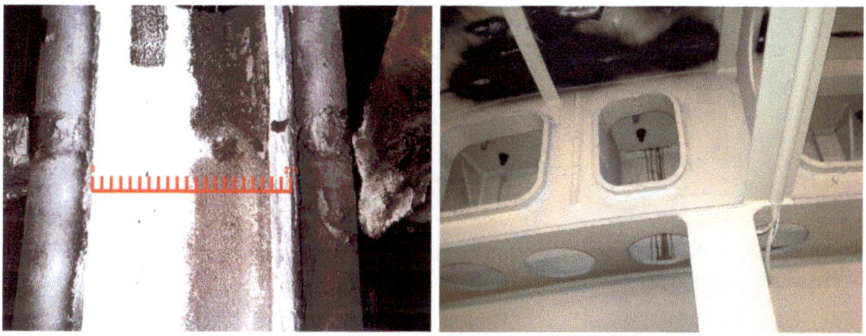

Fig. 2.19 Slewing wheel support of a bucket-wheel excavator: **a** lack of alignment in the contact between the bogie wheels and the raceway; **b** fatigue cracks in the slewing wheel support

Fig. 2.20 Examples of damage to the drive trains of tracked undercarriages

- Crawler crossbeam, multi-bogie-system—these are subject to typical fatigue processes, which involve fatigue cracks in high stress zones, such as axle bearings and tensile zones in the bottom sections of bogies. In terms of mechanical damage, bogies undergo damage resulting from overloads or material faults [8]. Examples of damage to these elements are shown in Fig. 2.21.
- Crawler pads—these are mainly subject to damage in the form of cracks in the area where subsequent pads are linked, resulting from torsional and bending loads acting on the links of individual pads. Additionally, although less frequently, cracks occur in load-carrying areas of pads on which wheels travel. Examples of damage to these elements are shown in Fig. 2.22.

(a) **(b)**

Fig. 2.21 Examples of damage: **a** fractures of the lower flange of a spreader's bogie; **b** damaged bogie wheel in the undercarriage of a bucket-wheel excavator

(a) **(b)**

Fig. 2.22 Examples of damage: **a** fracture of the crawler pad link; **b** damaged raceway of a crawler link

Tracked undercarriage steering system

This consists of the following elements, which are subject to operational problems:

- Actuator or steering shaft—the most popular three-point support system for surface mining machines comprises a fixed crawler unit and two steerable crawler units whose turn is controlled by a hydraulic actuator or a steering shaft attached to the end of the drawbar in steerable vehicles. During operation, this system is prone to problems typical for hydraulic systems, i.e. leaks resulting from accelerated wear caused by difficult working conditions in working levels in the mine. Damage also occurs to attachments of actuators and steering shafts to the undercarriage portal frame, which are caused by substantial loads transferred through this area, especially if the machine is operating on rough and wet ground. In winter, additional loads can be generated by the tracked undercarriage freezing to the ground.
- Steering and non-steering frames in the tracked undercarriage system—the load-carrying elements of the steering system also function as supports for the entire machine and, in addition to horizontal torsional forces, they also transfer vertical forces [9]. Damage to these elements appear in the areas of the greatest bending moments and shear forces, i.e. in the attachments of axles/half-axles in track frames, in the eyes of attachments between drawbars and the undercarriage portal frame, or in the areas of section changes and structural or technological notches. Damage also occurs to the above-mentioned half-axles of track frames. Nearly all of the damage to these elements is typically related to fatigue. Examples of damage are shown in Fig. 2.23.

Belt conveyors

The belt conveyor system of surface mining machines consists of the following main subassemblies, which are subject to operational problems:

- Belt conveyor drive system—this consists of single or, less frequently, double spur gears powered by electric motors. The gearbox is mounted on the shaft of the conveyor's drive drum, whereas its frame is supported like a typical moment beam on the other side by a supporting structure. During operation, this system is subject to damage typical of spur gears, resulting from natural wear processes. An additional factor that contributes to gear damage are start-up and stop overloads when the boom with the belt conveyor is tilted. This causes an additional dynamic surplus, which, in the case of lacking or incorrect coupling and/or brake adjustments, may even lead to shear damage to the system. There is also damage within the drive drum in the form of damage to the rigid connection between the gear and the drum shaft, as well as damage to the drum itself in the form of fractures in the connection between the side wall of the drum and the shaft. Fractures also occur in the drum shell, which are exacerbated by the pendulum-like cyclic bending of the drum unit with a very high number of cycles. Nearly all of the damage to these elements is typically related to fatigue. Examples are shown in Fig. 2.24.

Fig. 2.23 Examples of damage: **a** damaged end of crawler frame half-axle; **b** end of the half-axle before damage; **c** fracture of the spherical cup of the supporting ball in the frame support of a spreader's tracked undercarriage; **d** fracture in the attachment area of the excavator's non-steering frame

Fig. 2.24 Examples of damages—fracture of the drive drum of a bucket-wheel excavator's belt conveyor

- System of belts and rollers—consists of a belt and the supporting rollers. At the points of transition between conveyors, there is an increased number of rollers, whose function is to absorb some of the energy generated by the falling material. In such areas, the belt most often gets damaged in the form of rips, cuts, punctures or fractures. The dynamics of the phenomenon also cause damage to the rollers and their mounts as well as to the supporting structure. Similar damage to rollers occurs along the entire length of the belt conveyors, but is less severe. The heterogeneity of the transported material, especially the overburden cut by excavators, in the form of chunks, and sometimes large stones, accelerates the process of wear and the frequency of damage to the conveyor belt systems of surface mining machines. Examples are shown in Fig. 2.25.

Load-carrying systems of steel structures in surface mining machines
The load-carrying structures of surface mining machines consist of the following main elements, which are subject to numerous operational problems:

- System of load-carrying structures of booms:
  - Bucket-wheel boom—this load-carrying structure experiences the highest effort in terms of strength of materials. It is subject to direct impact of

Fig. 2.25 Examples of damage: **a**, **b** damage to conveyor belts; **c** damage to elements that support the rollers

variable cutting forces, i.e. the digging and side forces. The dynamic phenomena that accompany such forces, i.e. overloads to the excavation unit, intensify the damage that mainly takes the form of fatigue cracks in welded joints or in the native material, in geometric notches. Most of the damage is located on the boom head and in the points where it is attached to the excavator platform, where large values of bending moments occur. Damage also appears in the truss joints of the boom's horizontal and vertical sections. Shear damage may also occur to beams and structural joints when the boom collides with external obstacles. Examples of damage to excavating booms of bucket-wheel excavator are shown in Fig. 2.26.

– Feeding and discharge booms—these are elements of both excavators and spreaders. Main loads to these systems are generated by the overburden, the tilt of the machine and the dynamic horizontal and vertical loads that occur mostly during excavator or spreader travel [10] [11]. Damage occurs rarely and mainly takes the form of fatigue cracks in welded joints or in the native material, in the vicinity of geometric notches. Damaged areas are difficult to determine due to their small number.

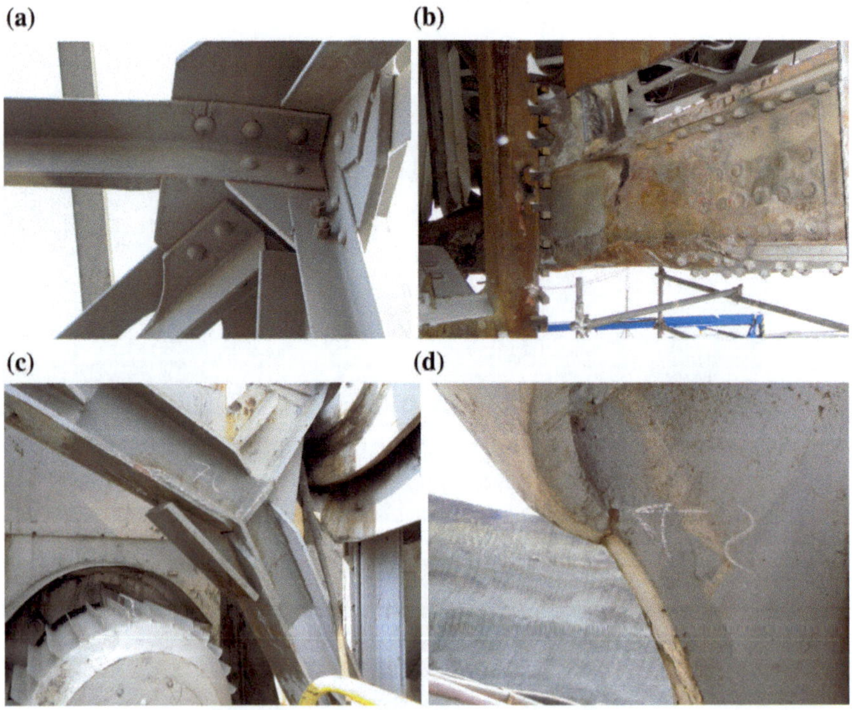

Fig. 2.26 Examples of damage to the bucket-wheel boom: **a** deformation and fracture of structural joint; **b** ultimate failure to the horizontal beam of the boom, near the bucket wheel; **c** fracture of the joint near the attachment of the torque beam of the bucket-wheel drive, **d** fracture in the eye reinforcement of the boom attachment

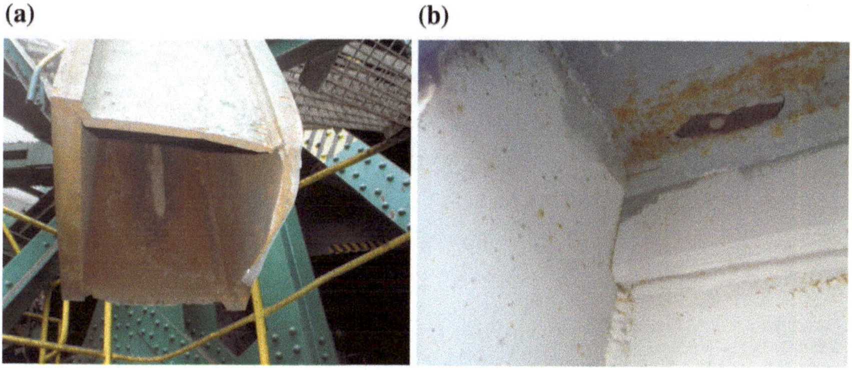

Fig. 2.27 Examples of damage to counterweight booms: **a** fatigue cracks of the counterweight boom tie rod [13]; **b** fracture of the top girder of the main counterweight boom on a bucket-wheel excavator

- Counterweight boom—this is an element of both excavators and spreaders. It provides stability and, by means of a winch that is attached to it, it performs the hoisting function of the excavating and discharge booms. Main loads are generated by the tilt of the machine and the dynamic horizontal and vertical loads that occur mostly during excavator or spreader travel or during digging of the excavator. Damage mainly takes the form of fatigue cracks in welded joints or in the native material, in geometric notches [12, 13]. Damage occurs in the joints of vertical and horizontal bracing and in suspension systems. Examples of damage to counterweight booms are shown in Fig. 2.27.

- Masts—these are elements that function as support for the booms of surface mining machines. Depending on the type of the machine, there can be single or double masts. Main loads to these systems are generated by the tilt of the machine and the dynamic horizontal and vertical loads that occur mostly during excavator or spreader travel. Damage occurs rarely and mainly takes the form of fatigue cracks in welded joints or in the native material, in geometric notches. Damaged areas are difficult to determine due to their small number.
- Pylon—this central part of surface mining machines is supported by the superstructure platform, to which the excavating and counterweight booms are attached in medium-sized and compact bucket-wheel excavators. In this case, it is most often a plate girder. In the case of spreaders, pylons typically take the form of 3D frames attached to the superstructure platform and are connected to the counterweight boom. Attached to them are discharge booms and masts. Damage mainly takes the form of fatigue cracks in welded joints or in the native material, in geometric notches. Damaged areas include the connections with the superstructure platform, structural joints of the plate girder and the frame. Damage also occurs near the connection with the counterweight boom. Examples of damages to pylons are shown in Fig. 2.28.

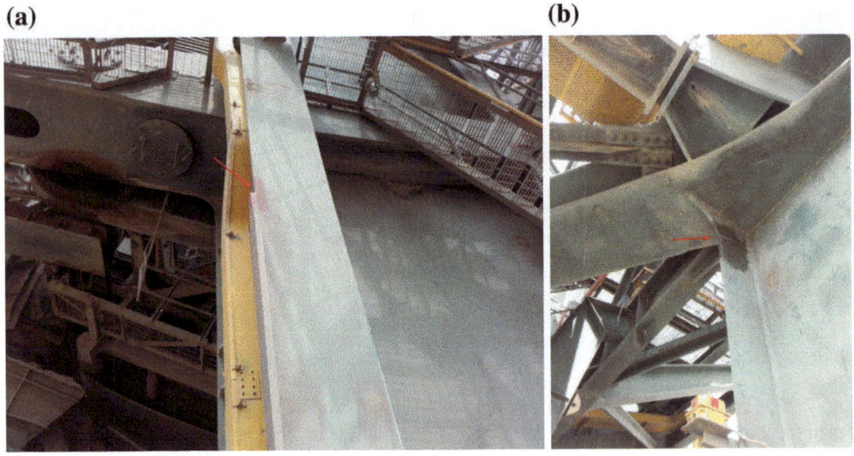

Fig. 2.28 Examples of damage to pylons of surface mining machines: a damage to the vertical pillar of the pylon; b fracture in the structural joint

- Slewing platform—the superstructure slewing system of surface mining machines, which supports other elements of the superstructure. It is supported in its lower part by a slewing bearing or slewing wheel support, and then by the undercarriage portal frame. Usually, it has the form of a plate girder connected to the pylon in the case of medium-sized and small machines. In the case of large machines, it supports the excavating and counterweight booms. Damage to the slewing platform primarily consists of fatigue cracks in welded joints and the native material located mainly near the connection with the pylon or with the booms. Fractures also occur in the attachment of the superstructure slew bearing or slewing wheel support bogies. Examples of damage to slewing platforms are shown in Fig. 2.29.
- Undercarriage portal frame—this is a load-carrying structure that supports the rotating part of the superstructure in surface mining machines, by means of a slew bearing or a slewing wheel support. It is usually a plate girder supported at the bottom by crawler units. Damage to girders is located mainly near the joint of superstructure rotation, as well as in the connection between the ring and the supports, which typically form a triangle. Damage also occurs around the connection with crawler units. In most cases, damage is related to fatigue, both in the case of welds and the native material. Examples of damage are shown in Fig. 2.30.

(a) (b)

Fig. 2.29 Examples of damage to slewing platforms of surface mining machines: **a** fractures in the platform diaphragm; **b** fracture in the connection between the pylon and the slewing platform

(a) (b)

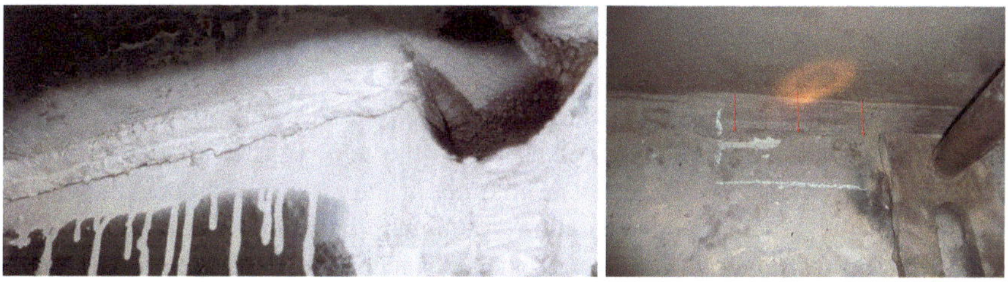

Fig. 2.30 Examples of damage to the undercarriage portal frame of surface mining machines: **a** fracture of the reinforcing rib in the connection between the web and the *top* flange of the frame; **b** fracture of the vertical diaphragm in the welded joint

2.4 Main Phenomena During the Operation of Surface Mining Machines

The phenomena that generate operational loads to surface mining machines also determine the frequency and character of their potential damage. Such phenomena can be generally divided as follows:

- Those that occur during excavation (digging forces, dynamic phenomena, loads in kinematic nodes),
- Those that occur during operational movements (primarily during machine travel).

Their influence on the individual elements of surface mining machines is different and depends on the range and intensity of interactions. The following section discusses the individual loads generated during the operation of surface mining machines.

Loads generated during excavation

The two main forces acting on the load-carrying structure of bucket-wheel and bucket-chain excavators during operation are as follows:

- Digging force U: this acts on the cutting edges of the bucket wheel tangentially to the wheel, with its orientation resulting from the direction of wheel rotation and from the value that is mainly the result of the amount of power transferred to the wheel shaft by the installed drive system, consisting of an engine and a drive gear,
- Side force S: this acts on the cutting edges of the bucket wheel in the horizontal plane and perpendicularly to the plane of the bucket wheel, with its orientation resulting from the current direction of excavation and from the value that is mainly the result of the amount of power expended by the relevant mechanism to rotate the machine in the horizontal plane.

Calculations of external loads acting on the bucket-wheel excavator [3] can be performed by means of two different methods. The first is related to the use of standards that define loads based on the power of drives and overload factors [14]. This is the typical approach applied in the design of surface mining machines. The power of the drive, on the other hand, is determined on the basis of the total value of the impact of the material being excavated on the unit, i.e. the total cutting resistance F. In general, this value consists not only of the resistance associated with the breaking of the structure and separation of the material from the deposit, generally called cutting resistance F_s; but also of the resistance associated with friction of the extracted material against the walls of the unit, called friction resistance F_t; the resistance associated with horizontal movement of the extracted material by means of the mechanism, called movement resistance F_p; the resistance associated with filling the container with extracted material, called filling resistance F_n; or the resistance associated with lifting the dredged material within the tool, called lifting resistance F_H. Thus the total resistance can be generally presented in the following form [3]:

$$F_u = F_s + F_p + F_t + F_n + F_H + F_{odp} \tag{2.1}$$

where:

F_s—cutting force,
F_p—movement resistance force,
F_t—friction resistance force,
F_n—filling resistance force,
F_H—material lifting force,
F_{odp}—ground resistance force.

It should be noted that not every excavation process includes all of these resistance forces and that their share in the total resistance also varies. However, it has been determined that in the case of joint rock and consolidated dry material, the 70–90% cutting forces are predominant, whereas in the case of unconsolidated and loose materials, it is the friction and filling resistance or horizontal or vertical movement resistance that are predominant.

The second method for determining the forces that occur during operation is based on measurements of the actual excavation loads [3]. This approach is particularly desirable in technical condition assessment and in the prediction of the service life of existing machines, as it allows the actual load data to be taken into consideration when calculating residual life. This method employs dedicated and calibrated monitoring systems, usually based on strain gauge sensors.

A comparison of cutting forces obtained by means of the first and the second approach leads to the following conclusions:

- The average loads measured in normal operating conditions are consistent with the values calculated on the basis of standards,
- The measured maximum loads that occur, especially during impact phenomena when the bucket wheel of the excavator suddenly stops, are several times larger than the values calculated using the standards. Such phenomena occur during excavation of overburden with non-mineable formations, such as rocks or stones, which is shown in the examples in Figs. 2.31 and 2.32.

Fig. 2.31 Overburden with non-mineable formations in a surface mine

Fig. 2.32 Broken bucket corner of a bucket-wheel excavator after a collision with a boulder

Loads occurring during operational movements

The operational movements of surface mining machines, i.e. travelling over the working level and all movements of the individual booms, as well as rotation of the superstructure, generate dynamic loads that affect the load-carrying structures of these machines. Experimental tests confirm that machine travel plays the most important part in the generation of dynamic phenomena. Such operational movement excites broadband vibrations of the entire machine and its individual elements. In such cases, global and local vibrations are excited. The values of these loads are established in standards as dynamic factors. They are also diversified depending on the type and element of the surface mining machine (Tables 2.1 and 2.2).

The values of dynamic loads can also be measured on actual machines. This is carried out by means of data monitoring systems with acceleration sensors located on the machine under test. Through this procedure, time traces and amplitude-frequency spectra of vibration acceleration values are obtained, which then serve as the basis for calculating the actual dynamic factors acting on individual elements of surface mining machines. The vibration sensor installed on the mast of a SRs-2000 excavator is shown in Fig. 2.33.

Examples of time traces registered during the operation of a bucket-wheel excavator are shown in Fig. 2.34 and its spectrum is shown in Fig. 2.35.

Table 2.1 Values of dynamic factors according to DIN-22261 [14]

Machine type	Machine element	Dynamic effects factors		
		Vertical D_V	Transverse D_Q	Longitudinal D_L
Bucket wheel excavator	Bucket-wheel boom	1/10	1/60	0
	Tower or central structure with counterweight boom	1/25	1/30	0
Bucket-chain excavator	Bucket-chain boom with cable supports	1/7	1/30	0
	Main support frame in the hoisting area of the chain boom	1/10	1/30	0
	Central structure	1/30	0	0
	Counterweight boom	1/20	1/50	0
Crawler-mounted spreaders	Discharge boom	1/10	1/10	0
	Tower—central part	0	0	0
Crawler-mounted machines	Counterweight boom	1/20	1/15	0
	Connecting bridges	1/20	1/10	1/15
	Cabs for operators	1/2	1/2	1/2

Table 2.2 Values of dynamic factors AS-4324.1 [15]

Machine	Machine part	Dynamic effects factors		
		Vertical	Horizontal	
			Transverse	Longitudinal
All rail mounted machines	Without digging element	1/10	1/30	1/30
	With digging element	1/8	1/30	1/30
Crawler-mounted machines and equipment with mechanical or hydraulic lifting feet	Bucket wheel boom	1/5	1/30	1/30
	Discharge boom	1/5	1/10	1/30
	Counterweight boom	1/5	1/15	1/30
	Tower or central structure	1/5	1/30	1/30
	Connecting bridges	1/5	1/10	1/15
All machines	Cabs for operators	1/2	1/2	1/2

2.5 Disadvantages of Traditional Methods of Dimensioning Surface Mining Machines

Long-term studies of surface mining machines, including both numerical methods and experimental tests, based on conditioning and data monitoring systems that use vibration sensors and strain gauges, make it possible to assess and identify the disadvantages of traditional dimensioning methods for this type of machine. This assessment addresses two main issues:

Fig. 2.33 Vibration sensor installed on the mast of a SRs-2000 excavator

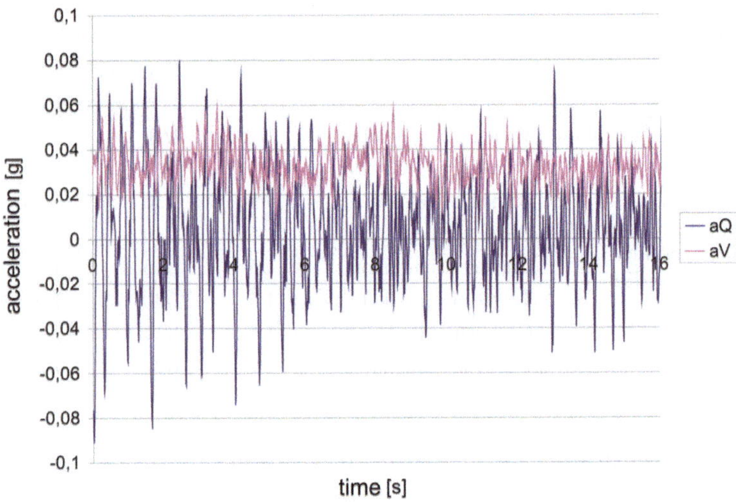

Fig. 2.34 Examples of time traces of horizontal (aQ) and vertical (aV) vibration accelerations measured on the SchRs 4600 excavator

- The incorporation of proper values of operational loads in calculations,
- The accuracy of determination of strength parameters of load-carrying structures in calculations.

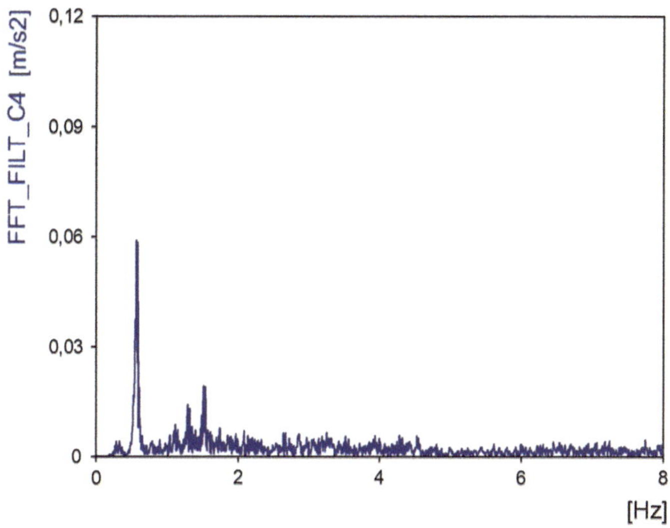

Fig. 2.35 Amplitude-frequency spectrum of vibration accelerations measured on the SchRs 4600 excavator

With respect to operational loads, several years of research into surface mining machines has led to an important conclusion that, in many cases, the standard approach to designing surface mining machines in which vibrations are accounted for by means of factors that increase constant loads, results in an underestimation of such loads and therefore in an incorrect estimation of durability. This situation pertains to, among others, excavating units of surface mining machines, especially those that strip overburden, in whose case the loads are determined by variable cutting resistances and dynamic phenomena generated by non-mineable formations in the material being excavated. During operation, excavating units frequently experience overloads that are not included in standard fatigue calculations and that may significantly influence the pace of the wear process. Figure 2.36 shows an example of a time trace of the digging force registered on a bucket-wheel excavator

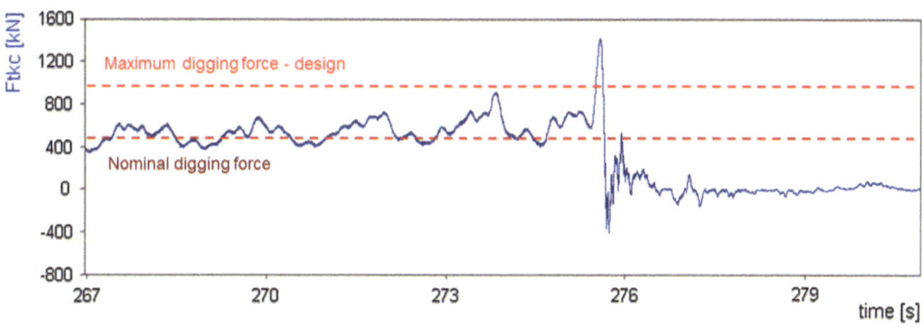

Fig. 2.36 Overload to the excavating unit and load values permitted by design

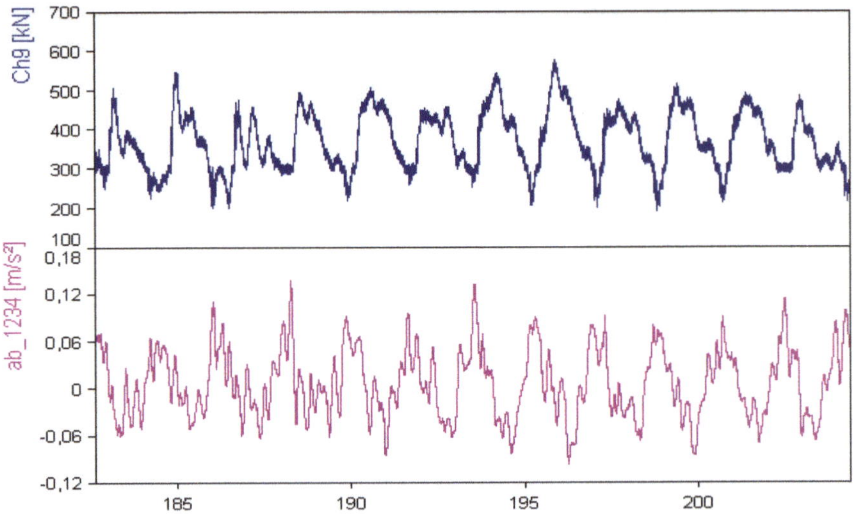

Fig. 2.37 Time-trace of changes in the digging force (*top chart*) and a similar time-trace of vertical vibrations of the excavating boom

with the indicated overload to the unit. Figure 2.36 also indicates the maximum load value permitted by the design, which is much lower than the actual value.

Figure 2.37 shows a similar time-trace of the dynamic effects that occur during excavation of the bucket-wheel excavator. There is a visible impact of vertical vibrations of the bucket-wheel boom on the value of the digging force U. These vibrations cause an additional amplitude of the cutting force, whose value is approximately 50% of the nominal force. This obviously affects the load to the load-carrying structure of the excavator and decreases its service life. These phenomena, however, are not included in traditional or standard-based methods of dimensioning surface mining machines.

Dynamic phenomena in the form of vibrations of surface mining machines and their individual components occur continuously throughout their service life. Vibration amplitudes described in the dimensioning standards for surface mining machines by dynamic factors are often underestimated in relation to the actual measured values. This has a direct influence on the underestimation of such loads in the process of fatigue development of load-carrying structures and thus can lead to fatigue damage.

Experience resulting from tests carried out so far on surface mining machines has led to the following conclusions:

- In many cases, the values of dynamic loads that act on load-carrying structures of surface mining machines exceed the values set or calculated on the basis of relevant standards.

- There is a correlation between the structure type of surface mining machines (e.g., C-frame, compact and large machine) and the nature and level of global and local vibrations.
- The effect of dynamic loads on the the decrease in durability is the greatest for counterweight and discharge booms. Subsequent elements of the bucket-wheel excavator's load-carrying structure towards the bucket-wheel boom are also subject to dynamic loads, but their influence on the generation of effort and fatigue damage decreases.
- With the increase in size of bucket-wheel excavators, the multiple of the overload factor in the bucket-wheel drivetrain decreases. Similarly, excavators with lower capacities and mass generate larger overload values that affect the that result in the decrease the durability.

The accuracy in calculations of strength parameters of load-carrying structures directly influences the quality of all calculations related to the development of load-carrying structures of surface mining machines. This primarily pertains to calculations related to ultimate and fatigue strength. Dimensioning standards for surface mining machines are based on analytic or numeric calculations but they use simplified models (beam or rod models). This approach necessitates increasing the values of partial safety factors. By doing so, an appropriate reserve resulting from the estimation of the level of stress in structural nodes can be taken into consideration in computations. This results in machine designs that are not optimal (heavy), but may also lead to a faulty design of load-carrying structures and result in their damage. What could help in such situations is the use of detailed shell or solid numerical models that represent the geometric characteristics of structural nodes and thus allow for a much more accurate determination of their stress state. There is a tendency to increase the accuracy of calculations and the designs of load-carrying structures, including those of surface mining machines, which improves the quality of these processes.

Given the above information, it seems justified to thoroughly update standard guidelines for designing surface mining machines. Such changes must be made both to the aspects regarding the determination of operational loads, as well as the adjustment of standard levels of safety factors and calculation methods to the currently used tools in this regard. It is also important to subject crawler undercarriages of surface mining machines to dimensioning requirements in terms of fatigue. Current guidelines only include calculations related to ultimate and buckling strength, which is a serious error in the DIN-22261 standard [14].

References

1. Babiarz S, Dudek D (2007) Annals of failures and catastrophes of surface mining machines—in Poland. Wroclaw University of Technology Publishing House
2. Rusiński E, Moczko P (2002) Cause of the damage of the chassis girders of the bucket wheel excavator. Przegląd Mechaniczny 4'02

3. Rusiński E et al (2015) Technical state assessment of the surface mining machinery—in Polish. Wroclaw University of Technology Publishing House
4. Kasztelewicz Z (2012) Bucket wheel excavators and spreaders. Technology of operation, Art-Tekst, Kraków
5. Kowalczyk M, Rusiński E (2013) Adjustment of the characteristics of the overload couplings in the main drives units of bucket wheel excavators, Ebląg rack and pinion gears: theory and practice. In: Rybczyńska T, Łabanowski J. Elbląg, pp 15–31
6. Rusiński E, Cegiel L, Michalczyk A, Moczko P, Olejarz J, Pietrusiak D (2015) Investigation and modernization of buckets of surface mining machines. Eng Struct 90:29–37. doi:10.1016/j.engstruct.2015.02.009
7. Smolnicki T (2013) Large size machines rotation joints. Local and global phenomena; Wroclaw University of Technology Publishing House, Wrocław
8. Bosnjak S, Petkovic Z, Gnjatovic N, Milenovic I, Jerman B (2013) Impact of the track wheel axles on the strength of the bucket wheel excavator two-wheel boogie. Tehnički vjesnik 20 (5):803–810
9. Maślak P, Smolnicki T, Pietrusiak D (2013) Strain gauges measurements and FEM analysis of elements of chassis of open cast mining machines. Tehnicki vjesnik—Technical Gazette 20 (4):655–665
10. Czmochowski J (2008) Identification of modal models of excavators in lignite mining. Wroclaw University of Technology Publishing House
11. Rusiński E, Czmochowski J, Pietrusiak D (2012) Problems of steel construction modal models indentification. Eksploatacja i Niezawodność—Maintenance and Reliability 14(1):54–61
12. Bošnjak S, Zrnić N, Simonović A, Momčilović D (2009) Failure analysis of the end eye connection of the bucket wheel excavator portal tie-rod support. Eng Failure Anal 16(3):740–750
13. Rusiński E, Czmochowski J, Iluk A, Kowalczyk M (2010) An analysis of the causes of a BWE counterweight boom support fracture. Eng Failure Anal 1(17):179–191
14. DIN 22261-2 Bagger, Absetzer und Zusatzgeräte in Braunkohlentagebauen
15. Standards Australia (1995) AS4324.1: Mobile equipment for continuous handling of bulk materials Part 1—general requirements for the design of steel structures

Chapter 3
Methods of Condition Assessment

To ensure that mines are operated in a reliable and economically efficient manner, it is necessary to monitor the condition of surface mining machines that have been in use for a long time. However, machine diagnostics is not very popular in the case of such machines, despite the undeniable benefits of collecting data about their condition. The possible reason behind this state of affairs is that operators and maintenance engineers have insufficient experience and knowledge in this area. This lack of popularity somewhat contradicts the observation that the knowledge of the current condition of surface mining machines significantly improves the efficiency of an enterprise, especially since this condition changes over time and is the basis for making operational decisions. Owing to the available technology, it is possible to monitor the condition of such machines continuously or periodically. The scope and method of monitoring substantially depends on the initial condition of a machine and the adopted maintenance strategy. The three basic strategies of machine operation are as follows:

- Operation until failure occurs (corrective maintenance),
- Condition-based operation (preventive maintenance),
- Operation that prevents and correct failures (predictive maintenance).

The first two operation strategies are the most popular, but they are not necessarily very effective methods of ensuring proper machine condition. It is estimated that approximately 90% of surface mining machines are operated using these approaches. Given that these machines are crucial to the production process, the first strategy is obviously not optimal. It often leads to unplanned downtime, which significantly reduces production efficiency. If such a strategy is adopted, condition monitoring amounts to periodic inspections and attempts to identify the causes of faults that have occurred.

The other two strategies, in particular predictive maintenance, constitute a significant upgrade to machine operation. They are described in more detail in the following subchapters.

E. Rusiński et al., *Surface Mining Machines*,
DOI 10.1007/978-3-319-47792-3_3

An analysis of the character of typical defects in load-carrying structures of surface mining machines indicates that approximately 80–90% them are related to fatigue. For this reason, condition monitoring methods should aim to identify and predict such machine faults.

Continuous or periodic condition monitoring of machines is carried out using many methods and techniques. The most important include:

- Experimental methods and techniques—non-destructive testing, monitoring of operational loads and parameters, diagnostic signals (vibration, noise, etc.) and many others,
- Computational analytical methods,
- Computational numerical methods.

These methods and techniques are used to analyze the obtained information and, consequently, to determine the condition of the machine under test. Reliability of the condition assessment process is in this case closely related to the adopted methodology and scope of testing.

Based on over 30 years of experience researching surface mining machines, the authors have developed a comprehensive method of assessing the technical condition of such equipment. Moreover, this method can also be used to predict the service life of the machine based on its condition and current and future operating conditions.

The following subchapter describes the proposed method and its main components (experimental measurement methods and techniques and calculations), as well as the entire methodology.

3.1 Experimental Methods

Condition assessment involves the following experimental methods:

3.1.1. Non-destructive testing (NDT)—this is designed to detect existing fractures in elements of the machine's load-carrying structure, as well as corrosion defects or potential mechanical damage. NDT commonly includes visual inspection, magnetic-particle inspection or dye penetrant inspection. Figure 3.1 shows examples of fractures in the load-carrying structure of the SchRs800 bucket-wheel excavator.

3.1.2. Operational load monitoring—one of the most important aspects in monitoring and predicting the structural health of machines is the identification of real operational loads to be used in calculations. There are many studies, standards and guidelines as to what the values of these loads should be or how to determine them [1]. It can be assumed that the knowledge of the loads related to ultimate strength is sufficient to assess the condition, but problems emerge when measuring the spectra of loads, which determine fatigue strength. This problem is best solved by measuring actual loads on

Fig. 3.1 Examples of fractures in the undercarriage platform of the SchRs800 excavator, detected using non-destructive testing

the structure under test. Spectra of operational loads are identified by long-term or continuous monitoring of individual parameters. In the case of surface mining machines, bucket-wheel or bucket-chain excavators, the main external loads are generated by digging forces. These loads are identified by means of force sensors installed in the points where loads are transferred, by scaled strain gauge measurement systems or by indirectly measuring the energy parameters of drives (power consumption). One of the easiest and relatively accurate methods of identifying loads, which can be installed in existing machines, is the strain gauge measuring system. In the case of excavators, it is usually located on the elements of the bucket-wheel drive system, such as gear housing, drive shaft, torque beam or on load-carrying elements of the bucket-wheel boom or the slewing platform. Figure 3.2 shows an example of a point for measuring deformation located on a torque beam in the gearbox of the SchRs4600 excavator bucket-wheel drive. To enable such a system to be put into operation, it needs to be calibrated by applying a known force at the point of action of the real digging force. Figure 3.3 is an example of a time trace of the circular digging force recorded during operation. The graph shows a visible overload of the excavating unit, after which is stopped immediately. This system for measuring loads can be permanently installed on a machine in order to continuously monitor the loads acting on the excavating unit. In the case of periodic condition assessment, the system records loads for several weeks, in order to determine a representative spectrum of operational loads.

3.1.3. Another source of operational loads are forces generated by the machine's movements, such as travelling, starting of the drives, among others. Most often, these loads do not significantly affect the service life of the load-carrying structures in these machines. However, in cases where they are an essential component in the sum of all loads, they should be taken into

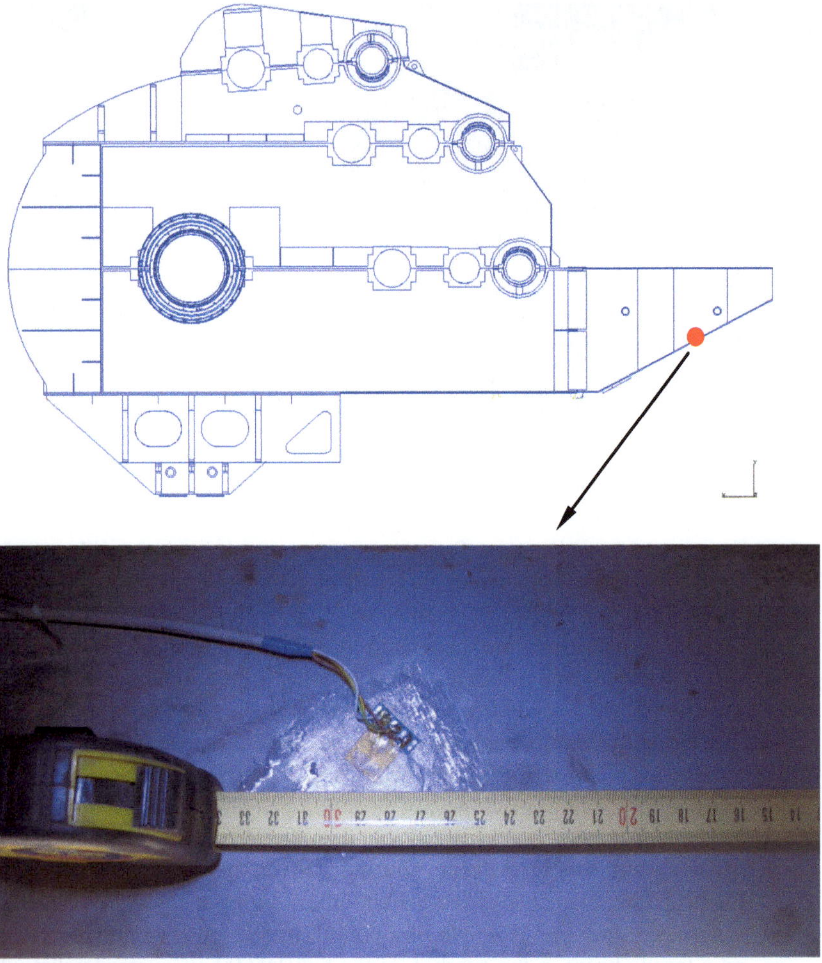

Fig. 3.2 Location of the point where deformation is measured on a torque beam in the gearbox of the SchRs4600 excavator bucket wheel drive

consideration during condition assessment. The approach to measuring these loads is similar to the aforementioned methods of measuring loads acting on the excavating unit. By using strain gauge systems, it is possible to create custom monitoring systems, which are adjusted to individual load cases that are to be identified.

3.1.4. Monitoring of other diagnostic signals—the findings of many years of research into surface mining machines indicate that the second significant parameter affecting the condition of load-carrying structures in these machines are vibrations generated by operational loads and the machine's movements. Therefore, the authors of this book adopted an approach that involves the monitoring of vibration in elements of load-carrying structures

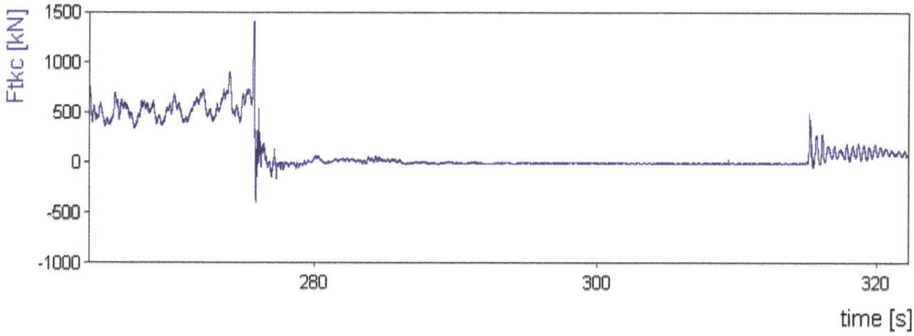

Fig. 3.3 Time trace of impact load recorded on the torque beam of the SchRs4600 excavator

and then using these results as a diagnostic indicator. In order to obtain information on the values of vibrations of individual elements in the machine's load-carrying structure, a monitoring system is required, which comprises low-frequency vibration sensors (accelerometers), cables and a data logging and analyzing system. Figure 3.4 is a diagram with examples of monitoring points on the SRs 2000 excavator. Figure 3.5 shows an example of a monitoring point located on the mast of the bucket wheel boom in the SRs 2000 excavator.

The results of monitoring include the spectra of vibration in monitoring points, which represent mode shapes of individual elements in the load-carrying structure of the machine under test. Figure 3.6 is an example of an amplitude-frequency spectrum for accelerations that represent the lateral vibration of the bucket-wheel

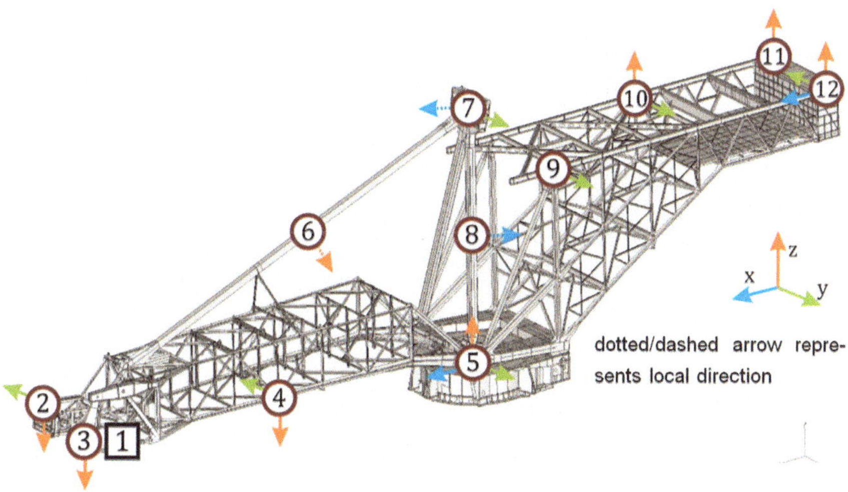

Fig. 3.4 Locations of monitoring points on the superstructure of the SRs 2000 excavator

Fig. 3.5 Example of a vibration monitoring point on the mast of the bucket-wheel boom in the SRs 2000 excavator

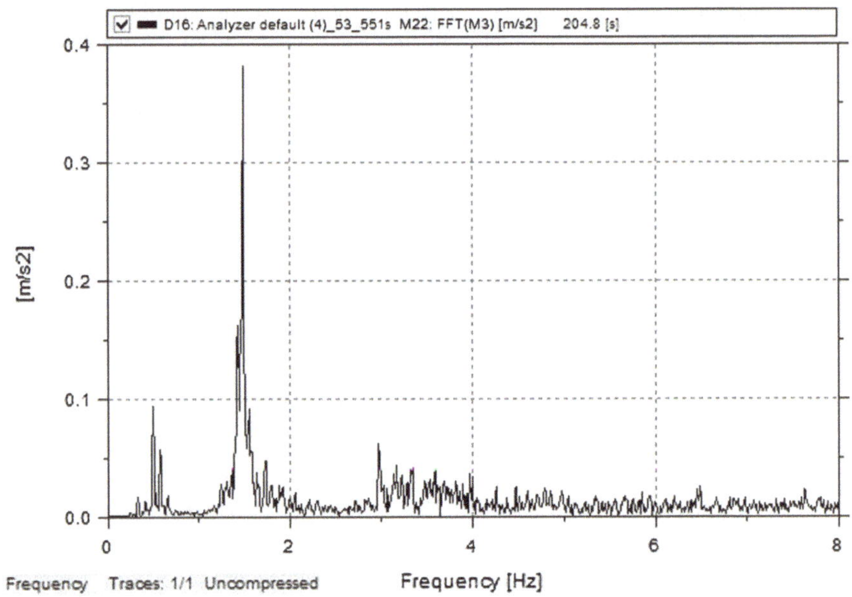

Fig. 3.6 Amplitude-frequency spectrum for accelerations—lateral vibration of the bucket-wheel boom

boom. Based on the time-traces, the effective RMS values of vibrations are calculated, which are then used as fatigue wear indicators and as values of dynamic loads in strength analyses on numerical models. The D indicator is calculated using the formula:

$$D = \frac{g}{\mathrm{RMS}}$$

where g is acceleration due to gravity, and the RMS value of vibrations [m/s^2] is determined on the basis of the measured vibrations.

Another factor related to the description of loads is the amount of low-cycle loads in the spectrum used for fatigue analysis. During operation, certain elements of devices (machines) are subject to overloads, which should not affect the operating safety of such structures. Designers usually predict such events but might underestimate their impact on the durability by assuming that their sporadic occurrence only affects the ultimate strength. This is related to the traditional method of designing, which involves fulfilling the ultimate strength requirement and fatigue strength requirement. Basically, these are two separate stages of the design process ([2] is a good example), which differ in the values of loads included in calculations. This approach can lead to the underestimation of durability and, consequently, to a failure of the machine.

3.2 Analytical and Numerical Methods

In order to assess the technical condition, it is necessary to define the state of stress using the identified, measured or standard loads. Traditional, analytical computational methods can be used to define the global state of stress of frames, or of flat or spatial lattice structures. However these methods cannot be used to precisely determine stress parameters in structural nodes and the local notches that usually occur in these nodes. The situation is more difficult for box structures with a complex geometry, which are usually used in surface mining machines. In this situation, analytical methods can only be used to estimate the state of global stress. This method is not sufficient to reliably assess and predict the technical condition. Numerical methods are thus used to identify the spatial state of stress of the investigated structures. This is usually performed using the finite element method (FEM) or, on rare occasions, the boundary element method or the finite difference method. Depending on the studied approach used to define the fatigue process (high-cycle or low-cycle), non-linear or linear calculations can be carried out with respect to materials.

Modeling of geometry and determining its effect on the state of stress using numerical methods (FEM, BEM, FDM) seems to be the best choice for detecting potential areas of fatigue cracks. In the case of existing structures, it is much more difficult, or even impossible (e.g., in joints), to determine the state of stress directly in the notches using experimental tests. The aforementioned numerical methods solve this problem in most cases. Depending on the adopted approach to defining fatigue, such calculations may either be performed for a model of material that is close to the real one, or by assuming that the model is ideally elastic. The latter approach is suitable for high-cycle fatigue analysis. The numerical model has one more significant advantage. It can also take into account geometric changes that were not planned by the designer and that resulted, for instance, during technological processes. Such changes can include manufacturing flaws, deviations from designed radiuses of curves or misalignments, as well as angle deviations of joined elements, etc. If all these factors are taken into account, the quality of analyses improves and so does the final result, i.e. the assessment and prognosis of the durability, mainly through the identification of fatigue parameters.

3.3 Preventive and Predictive Approaches in Condition Assessment

The most reliable approaches to diagnosing the condition of surface mining machines and efficiently using them are the two abovementioned strategies, i.e. condition-based operation (preventive maintenance) and operation that prevents and correct failures (predictive maintenance).

Preventive maintenance—this strategy may be divided into two sub-categories:

- Operation with regularly scheduled renovations—this popular approach uses operational experience and technical information about the implemented equipment, which can be used to predict how long a machine or its element should be in operation before being replaced or repaired. During renovation, key elements of the machine, which are not scheduled for replacement or repair, are often inspected. If defects or excessive wear, for example, are found, a decision is made to perform corrective actions. This type of operation is relatively expensive, but ensures a certain regularity of repairs and expenditure to keep the machine in good health. It is the most popular approach in the operation of surface mining machines. If a failure occurs between renovation cycles, a repair is scheduled and decisions are made as to whether to alter the schedule of future renovations. In some situations, the scope of repairs is increased to also include renovation of key subassemblies. In such cases, it is possible to postpone the date of overhaul by a certain amount of time. If a critical failure occurs, or if such a failure occurs before the scheduled overhaul, the overhaul is usually delivered on the occasion of repairing the failure. The drawback of such operational strategy is unplanned downtime and increased costs resulting from unforeseen failures. Another problem is the inability to fully take advantage of the durability (bearing capacity) of individual elements of the surface mining machine, which are repaired/replaced during the overhaul. This is due to uneven wear of individual assemblies in relation to the scheduled renovation cycles.

- Operation based on the condition—this approach to operation requires periodic or, preferably, continuous monitoring of the condition of the structure under investigation. If symptoms of damage are detected, an immediate analysis of the situation follows and decisions are made to perform corrective actions. This method helps to detect failures early, but its efficacy is strictly determined by the quality of condition assessment. This pertains to the assessment methodology, the implemented monitoring techniques, computational methods and indicators of the process of degradation of the structure under study. Early detection of faults helps to reduce the costs of failure repair and decrease machine downtime. If it is not possible to perform a quick repair, a temporary correction allows the machine to continue operation until the next most convenient date of renovation. The machine can also continue operation on the condition that limitations are introduced; for example, lowering the thresholds of safety systems, limiting the output, changing the excavation site, and so on. Such actions help to limit the development of damage and offer flexibility in planning the methods, costs and, most importantly, the date of overhaul. An example of a basic action that delays repairs is stopping fatigue fracture from propagating by drilling a hole at the end of the fracture. Figure 3.7 shows an example of such a temporary correction.

Predictive maintenance—the fundamental difference between preventive and predictive maintenance is that the latter approach can predict the occurrence of

Fig. 3.7 Example of a method to prevent the development of fatigue fracture by drilling a hole at the end of the fracture

failure. Based on such information, users can take preventive steps in advance. These include activities related to both operation and repair. In terms of operation, it is possible to modify the parameters of the machine (similar to the case of operating with a defect, mentioned above), which can increase the operation time. When corrective actions are performed in advance, structural modifications can be introduced in areas where the fault is predicted to occur, which will decrease the risk of failure. Such actions may be performed at any moment; for example, during downtime or planned overhaul. The purpose of both of these categories of activities is to postpone, or eliminate entirely, the expected failure. If repairs are required, or if a faulty element must be replaced, preventive actions make it possible to order new machine elements well in advance of the overhaul. In some cases, it is also necessary to make an early decision to pull the machine out of service in order to prevent a serious or catastrophic failure.

However, the predictive maintenance strategy requires the application of the most advanced methods of condition monitoring. Depending on the machine, it is usually necessary to perform a wide range of experimental tests, implement advanced measurement techniques and apply complicated computational methods to predict the machine's behavior. The efficiency of this approach to operation strictly depends on a wide range of tools and is described by the probability of accurately identifying the location and time of failure. This efficiency is usually higher with respect to locating the potential failure site, whereas the moment of failure is much more difficult to predict. This is especially true of fatigue damage.

Among the currently used methods of condition monitoring, which help to oversee the operation of various types of machines, of higher popularity are those that enable the identification of damage to the machine resulting from a long-term process of degradation [3–5]. As for the operation of surface mining machines, there are no comprehensive methods that would accurately assess the technical condition and, most importantly, predict the occurrence of a failure and thus the residual life. For this reason, the authors have proposed and implemented their own numerical-experimental method for assessing the technical condition of surface mining machines, which can predict residual life of these structures after long-term use. This method is described in the following chapter.

3.4 Combined, Comprehensive Experimental and Numerical Condition Assessment Method

The basic question related to the operation of different types of machines and equipment, including surface mining machines, is how much longer can they operate safely after a certain number of years in service. This is primarily the problem of residual life. Such questions are difficult to answer, especially since there are no guidelines that would align the procedure in such cases. The situation is similar in the case of surface mining machines, which are very expensive to build and removing them from service can be problematic. On the other hand, they must ensure the safety and reliability of operation, and these two parameters are hard to predict for machines that have been in service for decades. The main problem in this case is the load-carrying structure of these machines, which is hardly ever renovated, in contrast to the mechanical equipment. This is the reason behind the proposed numerical-experimental method for assessing the technical condition and predicting residual life of the load-carrying structures in machines. Figure 3.8 shows a block diagram of the method. Its key component is the identification of the loading state, which is performed mainly by means of experimental tests. The main focus of this method is on the numerical identification of the state of stress using 3D computational models based on the FEM, which are then used for fatigue calculations. The next activity, which determines the reliability of this method, is the prediction of residual life based on the identified state of stress and loads in the structures under study. The final, complementary step in the proposed method is non-destructive testing aimed at determining the condition of the load-carrying structure, especially in areas directly responsible for excavator safety and in sites identified as highly stressed, using fatigue calculations. These tests complete the knowledge of the degree of degradation of the load-carrying structure and its estimated service life.

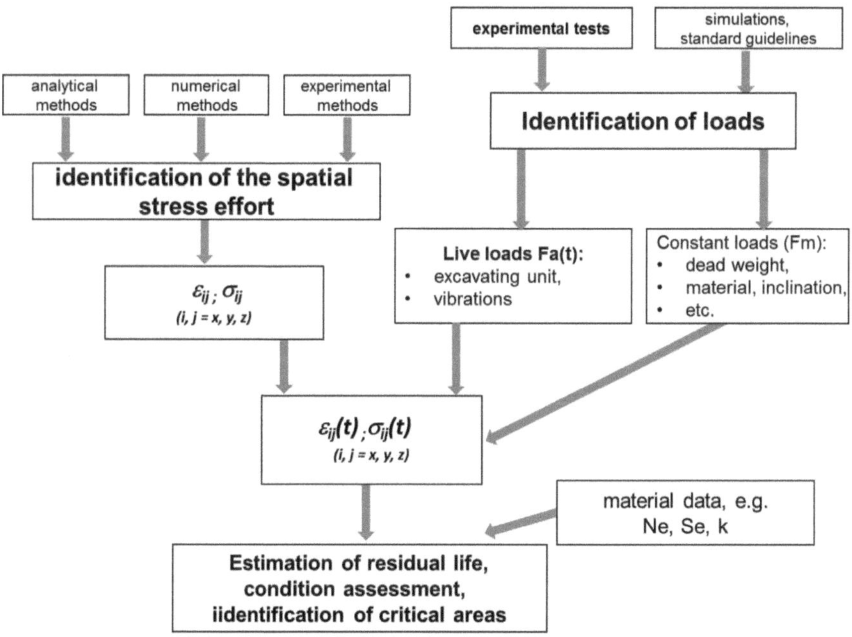

Fig. 3.8 Block diagram for the method of assessing the condition of surface mining machines and predicting their residual life

3.4.1 Identification of Loads

A very important element of the proposed method is the load spectrum. It is usually identified based on experimental tests. If such tests cannot be conducted, one can use standard guidelines or analytical calculations based on mechanical and physical equations. Calculating operational loads is difficult, especially if loads are stochastic in nature. A change in loads over time depends on the conditions of use, which are often variable due to output requirements or soil structure variation. Depending on the function of the machine, the randomness of loads may vary. This is often the case during the operation of surface mining machines, particularly overburden excavators, whose loads are determined by variable digging forces and dynamic phenomena, generated by undiggable formations in the material being excavated. Figure 3.9 shows an example of a time trace of the circular excavating force measured on a bucket-wheel excavator. During machine operation, there are many dynamic overloads, whose values exceed theoretical maximum (design) values.

Figure 3.10 shows a similar time trace of dynamic effects that occur when individual buckets cut into the material being excavated. The measured signal pertains to the bucket-wheel shaft in the SchRs4000 excavator. The red lines represent the time trace with the number of buckets filtered out, whereas the black line represents the complete time-trace. Each impulse resulting from the bucket cutting into the material generates fatigue damage.

Surface mining machines are subject to variable dynamic loads and overloads. Therefore, the optimum methods for identifying the spectrum of loads for such structures are tests during operation (if feasible). In the proposed method, tests aimed at identifying loads related to surface mining machines pertain to the following types of loading:

- Loading of the excavating unit—circular and lateral cutting force,
- Dynamic loads (vibrations) of individual elements in the machine, generated during operation (excavating, machine's movements).

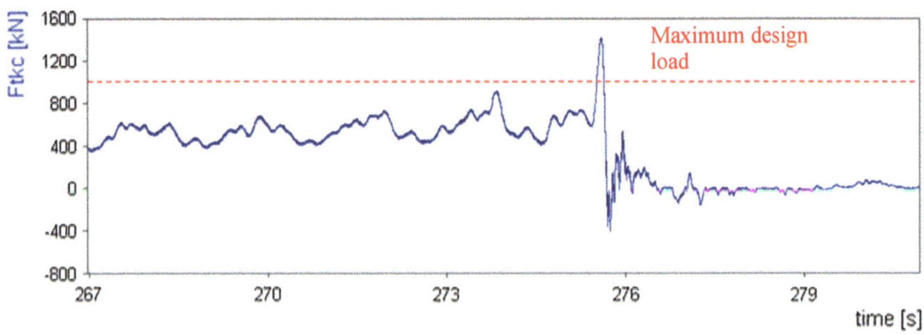

Fig. 3.9 Change in the circular excavating force over time, showing visibly exceeded theoretical design values

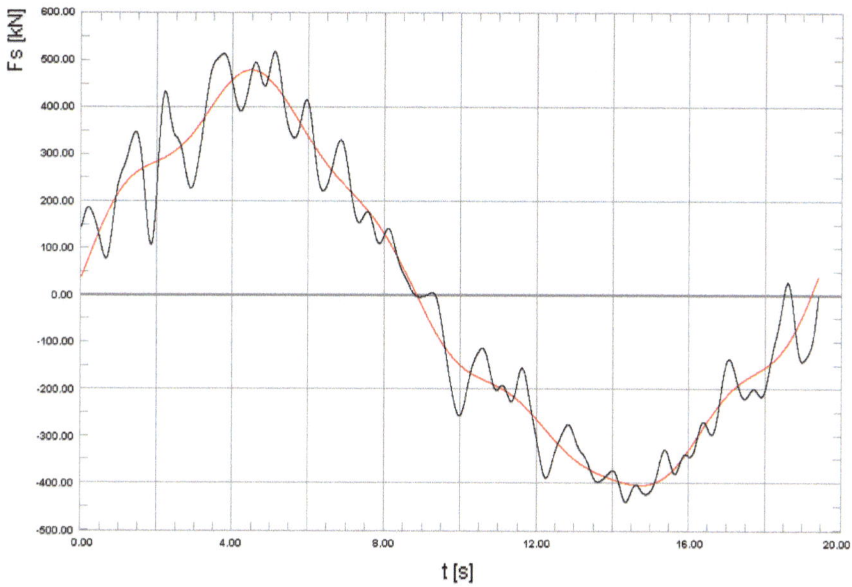

Fig. 3.10 Time trace of bucket-wheel shaft bending caused by the circular cutting force, F_s, with the components of buckets cutting into/leaving the material being excavated (marked *black*) and time-trace without such components (marked *red*)

The method of performing these tests is presented both in Sect. 3.1 and in the following examples of applications for this methodology.

These tests do not take into account the influence of other live loads, such as machine inclination, or presence of material on conveyors. Such loads are taken into account during fatigue calculations in accordance with standard guidelines.

Identification of loads may be performed on existing structures in limited time. It should be assumed that the shorter the monitoring time, the less accurate the test results. Therefore, selecting a representative load spectrum in this case requires long-term testing. However, if such tests cannot be conducted, it is common to use standard guidelines regarding the design of such devices and elements of machines. There is also the possibility to devise and implement special standardized time traces, as in the case of designing aerostructures [6]. In the case of surface mining machines, the load spectra measured on existing machines can be adjusted to other or new structures with similar structural characteristics. Such time traces are scaled and adjusted to the capabilities and technical characteristics of the structure under design. It is impossible, however, to include and predict all loads that can occur during operation. When identifying loads, especially if theoretical values are assumed (which are determined according to the standard), the level of uncertainty is compensated for by applying safety coefficients, which for surface mining machines have the value of 1.5. As a result of erroneously assumed character of loads (usually underestimation), the service life of the structures under design may be different (shorter) than initially assumed.

3.4.2 Identification of State of Stress

The next step in the proposed method involves the identification of the spatial state of stress, or field of deformations, in the structure under the loads identified in the previous step. This is usually performed by means of the FEM and, depending on the adopted approach to the fatigue process (high-cycle or low-cycle), the material calculations can be made with respect to a linear or non-linear model of material. The result of calculations is the spatial field of stress (deformations) in every point of the structure under investigation.

The fact that notches are included in calculations, by modeling their geometry, makes it possible to determine the complex spatial state of stress (deformations). The application of shell or solid modeling helps to obtain such representation and decrease the level of uncertainty of input data for durability analysis.

As a result of calculations, the spatial field of stress (deformations) is obtained for every point in the structure under test. Figure 3.11 is an example of a geometric representation of structural nodes in the KWK1500 bucket-wheel excavator. Apart from the 3D numerical model, Fig. 3.11 shows the calculated spatial field of stress and the real structure, which was constructed on the basis of simulations.

3.4.3 Methodology of Fatigue Calculations

After identifying loads acting on the structure and determining the state of stress in the element under study, its fatigue parameters are assessed. The proposed method offers two possibilities of estimating the durability of the load-carrying structure under test:

- A modified classic method of selecting the characteristics for the structural node, which includes a detailed geometric representation by means of 3D numerical models. This modification to the classic method of estimating durability is more accurate in identifying the state of stress by including geometric features in the model. This changes the status of an important parameter from an assumption to an accurate value (numerical methods such as FEM are approximation methods), which depends on the adopted modeling method. The following section discusses the method of identifying the state of stress by means of numerical models in more detail.
- Detailed analysis of durability of structural nodes taking into account their actual geometry, location of joints (e.g., welded, bolted joints) and spectra of real loads that cause cumulative fatigue damage. This approach, similar to the

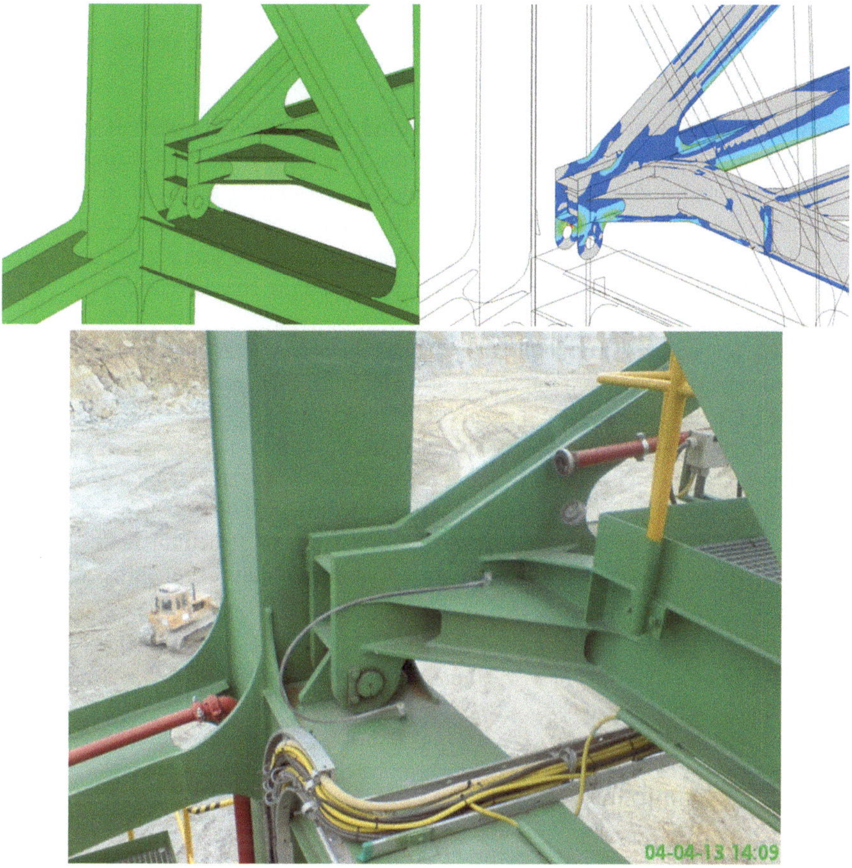

Fig. 3.11 Representation of geometry, contours of stress and view of the real structure—structural node of the KWK1500 excavator

modified classic approach, is based on a detailed identification of the state of stress by means of numerical methods. This is followed by definition of load spectra, which are used to analyze the fatigue process in terms of cumulative fatigue damage. The following section is a detailed description of the most important factors that determine the accuracy of this approach.

In the first method, after determining the state of stress in the element under investigation, a comparison of fatigue characteristics is made between typical structural elements and the nodes under design. In the case of surface mining machines, standard characteristics are used [1]. These characteristics differentiate fatigue strength according to the following parameters of the structural node:

- Parent material or weld,
- Type of welded joint,

- Quality of welded joint,
- Load state of welded joint,
- Geometry of structural node.

What is important in the proposed method is that the geometry of notches is included in the computational model of the structural node. This simplifies the calculations and makes them more accurate by eliminating the need to adjust the node under design to the standard geometric fatigue characteristics. Therefore, the use of standard characteristics pertains primarily to welded joints and data for the parent material. Figure 3.12 shows an example of fatigue characteristics of a welded T-joint.

Based on numerical calculations and fatigue characteristics, information is obtained regarding areas on the structure under study, where permissible values are exceeded for the assumed welding parameters. Figure 3.13 shows example results of such an analysis, where colors are used to represent contours of fatigue stress range in the slewing platform of a bucket-wheel excavator, which exceed the permissible threshold value required by the standard.

The second method of determining the durability of structural nodes is more complex. This procedure uses time traces of loads that were either measured

T-joints, (I joints) Structural shape and loads	Joint type	evaluation group	Perm. $\Delta\sigma\Delta\tau$	Code for K notch[1] case for table 24	Computational cross-section
22.7 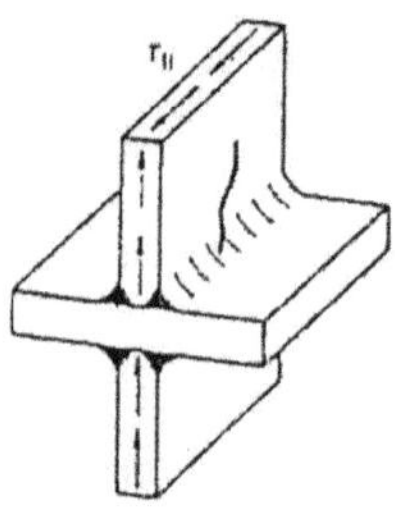 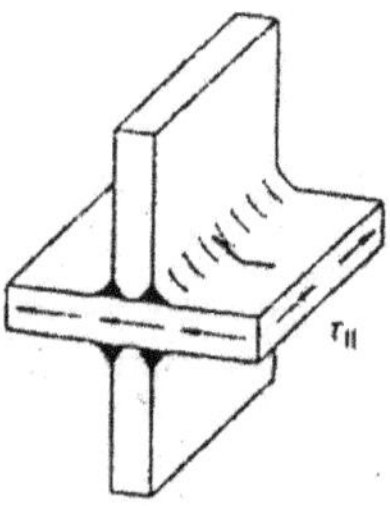	DBT	1 2 3 4	100 90 80 71	K_σ 6 7 8 9	Weld or structural element
	DET NBTc NBT NET	1 2 3 4	90 80 71 63	K_σ 7 8 9 10	

Fig. 3.12 Example of fatigue characteristics of a welded T-joint

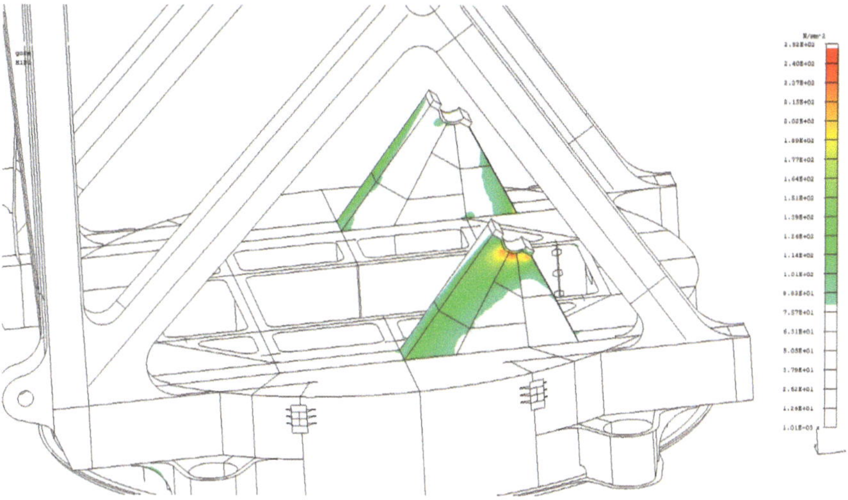

Fig. 3.13 Contours of fatigue stress range above the standard permissible threshold value [MPa] —slewing platform of bucket-wheel excavator

experimentally, numerically calculated, scaled, or are based on the standard. After determining the state of stress in the structural element under study, calculations are performed to find the value of fatigue damage caused by a single loading cycle or loading cycle block. Next, these damages are added (cumulated) and then the durability is calculated, which is expressed in the number of cycles until a fatigue fracture occurs. This procedure incorporates many hypotheses of cumulative fatigue damage. The most popular include:

- Palmgren-Miner rule. The damage function is described by the expression:

$$D_{\mathrm{PM}} = \sum_{i=1}^{q} \frac{n_i}{N_i} = 1 \tag{3.1}$$

where:

D_{PM} function of fatigue damage (failure) according to Palmgren-Miner hypothesis
n_i number of cycles with stress amplitude of σ_{ai}
N_i number of cycles to failure at stress amplitude of σ_{ai}
q number of stress levels.

The main drawback of this hypothesis is that the cumulative fatigue damage does not include stresses below the fatigue limit. Therefore, the following hypotheses are more useful for cumulative fatigue damage:

- Modified Palmgren-Miner rule—in this hypothesis, the S–N curve of limited fatigue strength is extended at the same angle, defined as *m*—cotangent of the S–N curve slope angle. This modification of the hypothesis includes the stress-inducing loads below the fatigue limit.
- Haibach hypothesis—the S–N curve of limited fatigue strength is extended at an angle defined by exponent m', which is related to exponent m in the following manner:

$$m' = 2m - 1$$

- Corten Dolan hypothesis, described by the dependency:

$$D_{\mathrm{CD}} = \sum_{i=1}^{q} \left(\frac{n_i}{N_i}\right)^{\rho(\sigma)} \tag{3.2}$$

where $\rho(\sigma)$ is the exponent, which depends on the stress at individual levels in the time-trace, and which usually has a value of (0.8–0.9) m.

- Serensen Kogajev hypothesis [7], where the damage function has the form:

$$D_S = \sum_{i=1}^{q} \frac{n_i}{N_i} = a_s \tag{3.3}$$

where:

$$a_s = \frac{\sigma_{\mathrm{amax}}\varsigma - kZ_g}{\sigma_{\mathrm{amax}} - kZ_g} \tag{3.4}$$

σ_{amax} largest amplitude in the time trace,
K coefficient describing the sensitivity of the hypothesis by determining the smallest value of stresses that may cause fatigue damage (this value is usually assumed at 0.5),
ζ duty cycle.

It is assumed that fatigue damage will occur when the sum *D* reaches the value of one. There are cases where this value is lower than one. The cycles and half-cycles of stress changes are usually counted using the rainflow method.

The results of this calculation procedure include areas in the load-carrying structure, whose fatigue damage value exceeds the threshold value (usually assumed as 1). Figure 3.14 is an example of result visualization for a structural node in the load-carrying structure. The efficiency of this method has been verified when a fatigue fracture was detected in this location on the real structure. This is also shown in this figure.

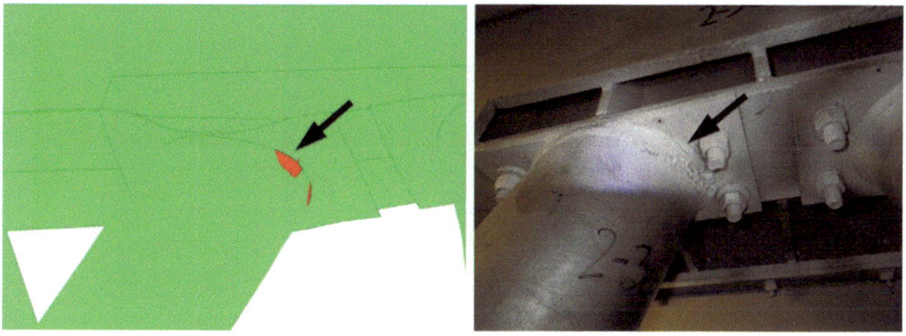

Fig. 3.14 Expected site of damage and its actual location in a structural node of the load-carrying structure of a jaw crusher

The proposed method of technical condition assessment and prediction of residual life of load-carrying structures in surface mining machines significantly contributes to the quality of the performed investigations. The use of modern computational numerical methods and experimental tests substantially clarifies the conducted analyses. It is a multi-purpose method, which can be used to predict and analyze the durability of load-carrying structures in all stages of the machine's life:

- Designing stage,
- Prototyping stage,
- Normal operation stage.

The method is sufficiently universal to be successfully applied to most load-carrying structures of machines and devices.

Examples of applications of this method are discussed in the following chapter.

3.5 Examples

The following section provides examples of applications for the method of assessing the technical condition health of surface mining machines, developed by the authors of this book.

3.5.1 Tests on the SRs 2000 Excavator

To assess the technical condition and predict the residual life of the SRs 2000 excavator, a modified version of the classic method was applied. This method is based on numerical models and real values of loads obtained through experimental tests, which were aimed at identifying the loads acting on the excavating unit and dynamic loads. To determine the state of stress of the load-carrying structure of the

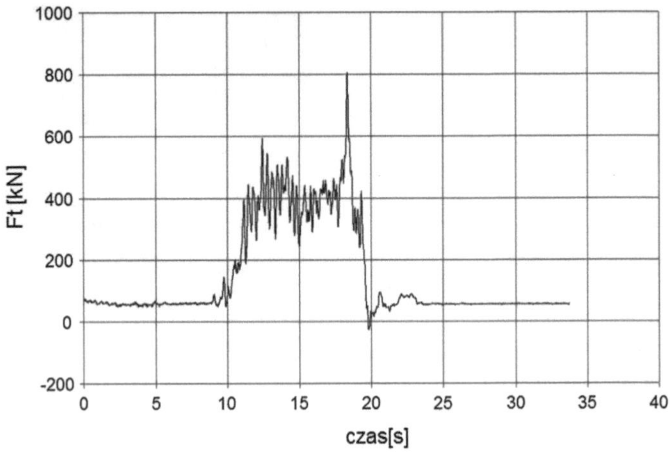

Fig. 3.15 Overload of the excavating unit under extreme load of circular cutting force of 807 kN

surface mining machine, numerical models based on the FEM were used. Fatigue load combinations were defined based on the standard [1].

First, loads were identified that act on the load-carrying structure of the machine in relation to:

- Loads to the excavating unit—Fig. 3.15 is an example of a time trace of the excavating unit under a load of circular cutting force, with a visible overload of the unit followed by the initiation of the safety system. Over the course of several weeks, loads to the excavating unit were monitored and the overload coefficient reached a maximum value of 1.55. Cases of overload exceeding 1.3 constitute approximately 1.5% of all overload cases exceeding the value of 1, and approximately 0.04% of operating time. The total time when the overload coefficient exceeded 1 was 10.75 h, which constitutes approximately 3% of the entire monitoring time. Based on the results of experimental tests, a nominal value of 520 kN for the circular digging force was assumed for numerical fatigue calculations, which represents the real operating conditions of the excavator under study.
- Dynamic loads—values of real dynamic coefficients resulting from vibrations of individual elements of the excavator's load-carrying structure, determined on the basis of experimental tests in accordance with the methodology discussed in Sect. 3.1 of this book. The values of these coefficients are presented in Table 3.1. Standard permissible values are noticeably exceeded (marked in bold) in all elements of the body except the mast. The mast, however, shows a high level of lateral vibration (X'), which is not included in the standard for fatigue calculations, and which in this case is a significant factor that can decrease the service life. Based on the obtained test results, the following real values of dynamic coefficients were assumed:

Table 3.1 Comparison of standard values of the dynamic load coefficient, *D*, with real values for the SRs 2000 excavator

Machine element	Point (Fig.3.4)	Direction	RMS [m/s²]	1/*D*	Minimum standard coefficient value		
					$1/D_v$	$1/D_q$	$1/D_l$
Bucket-wheel boom	2	*Y*	0.239	**40.99**		60	
		Z	0.317	30.95	10		
	3	*Z*	0.649	15.10	10		
	4	*Y*	0.202	**48.66**		60	
		Z	0.601	16.32	10		
	5	*X*	0.078	126.13			0
		Y	0.069	142.86		60	
		Z	0.042	233.47	10		
BWB tie member	6	*Y*	0.904	**10.86**	10		
Mast	7	*X*	0.142	68.97			0
		Y	0.263	37.32		30	
	8	*X'*	0.759	**12.93**			0
Counterweight boom	9	*Y*	0.182	53.86		30	
	10	*Y*	0.068	144.00		30	
		Z	0.134	73.48		25	
	11	*Z*	0.382	**25.69**		25	
	12	*X*	0.144	68.08			0
		Y	0.054	180.96		30	
		Z	0.347	**28.30**		25	

– For the bucket-wheel boom, mast and tie member:

$1/D_V = 10$
$1/D_L = 68$
$1/D_Q = 41$

– For the counterweight boom:

$1/D_V = 26$
$1/D_L = 68$
$1/D_Q = 30$

The identified values of real operational loads to the excavating unit and dynamic loads, as well as the standard values of other live loads (inclination, material handled on conveyors, friction forces at pivot joints, etc.) can be used to identify the state of stress of the excavator's load-carrying structure. To this end, a 3D numerical model of the excavator body was created, which is shown in Fig. 3.16. This shell model accurately represents the geometry of the existing structure.

Fig. 3.16 Numerical model of excavator SRs 2000 body

Based on the calculations related to fatigue strength, and taking into account real load values, a spatial state of stress is obtained for the load-carrying structure of the excavator under study, defined as the range of fatigue stress $\Delta\sigma$. Assuming the threshold value of permissible fatigue stress range, which is based on standard fatigue characteristics, areas on the structure under investigation can be identified where permissible values are exceeded for the assumed manufacturing parameters. Figure 3.17 shows an example of the results of such analysis, where colors are used to represent contours of fatigue stress range in the bucket-wheel boom of the bucket-wheel excavator, which exceed the standard permissible threshold value. The obtained calculation results are used to identify sites that are potentially prone to fatigue fractures. Figures 3.18 and 3.19 show structural nodes, welded joints, and areas of structural notches in the bucket wheel boom of the SRs 2000 excavator, which were identified using this method. In accordance with the developed method, these areas are then analyzed using defectoscopy in order to detect potential fatigue damage. If such damage is detected, the possibilities of repairing the node or, more often, modifying it are explored. In the final stage of the developed method, recommendations are made regarding periodic inspections of detected critical nodes in the load-carrying structure of the machine under investigation. Inspection intervals depend upon the level of stress obtained from calculations and the technical condition identified on the basis of defectoscopy.

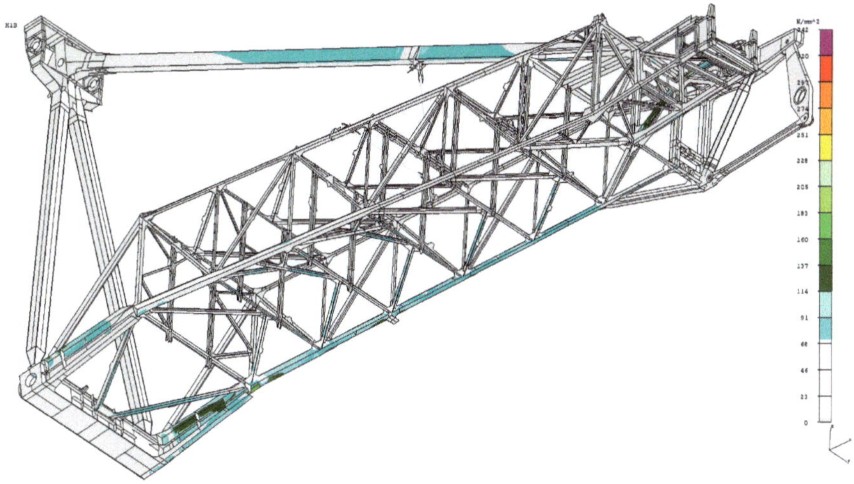

Fig. 3.17 Contours of fatigue stress range in the bucket-wheel boom of the SRs 2000 excavator [MPa]

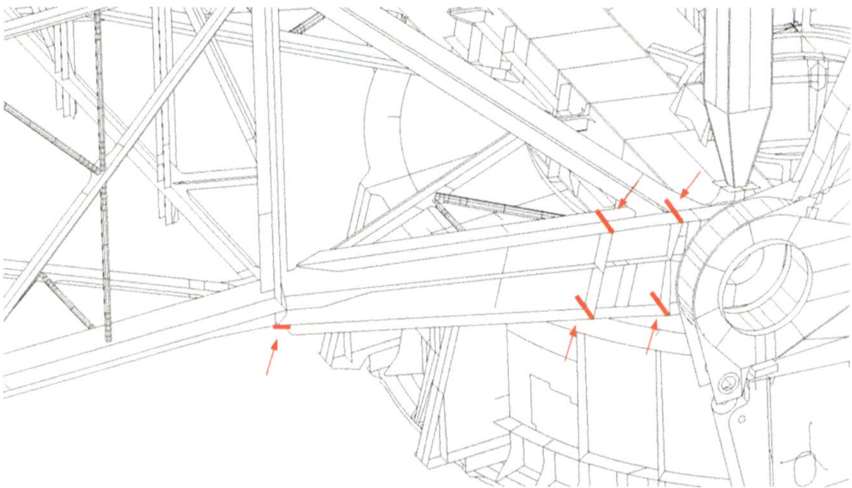

Fig. 3.18 Areas on the load-carrying structure of the bucket-wheel boom, which are potentially prone to fatigue damage and were scheduled for periodic inspection

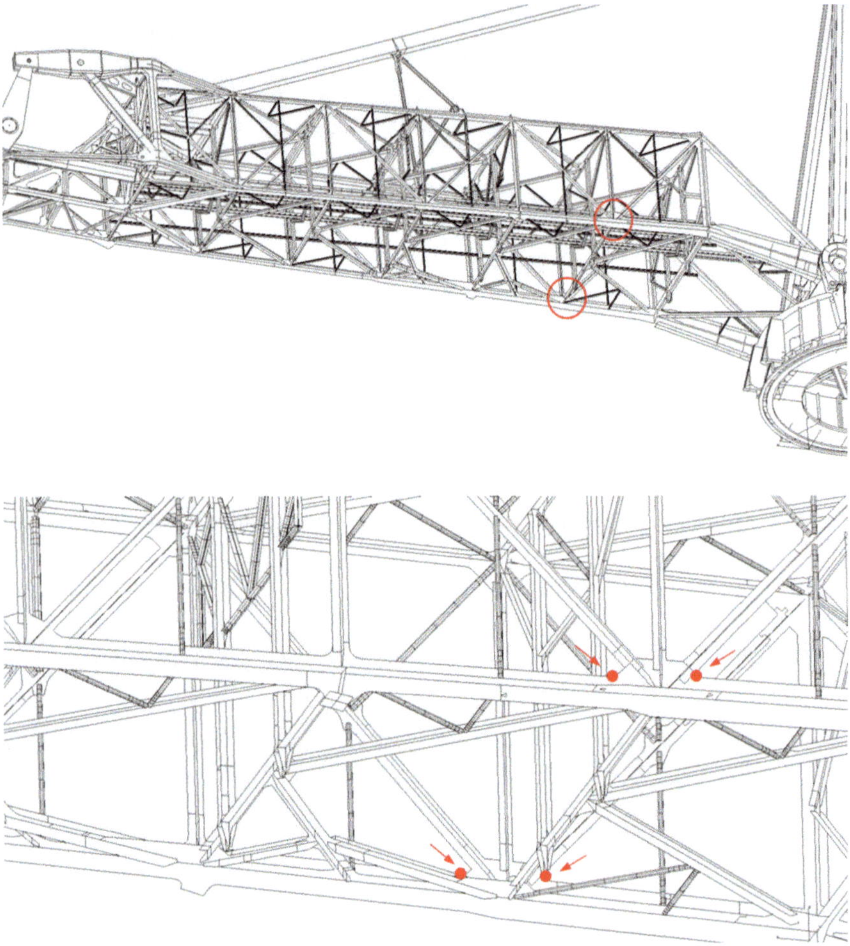

Fig. 3.19 Areas on the load-carrying structure of the bucket-wheel boom, which are potentially prone to fatigue damage and were scheduled for periodic inspection

3.5.2 Tests on the SchRs 4000 Excavator

The advanced method of predicting residual life, similar to the basic version, uses real values of loads acting on the excavating unit and of dynamic loads, which are obtained through experimental tests. Similarly, to determine the state of stress of the load-carrying structure of the surface mining machine, numerical models based on the FEM are used. These are usually shell or solid models. Fatigue combined loads are defined based on the standard, but most often loads of equal force are applied to the calculation model. The FEM is used in calculations to obtain the values of fatigue stress amplitudes in all areas of the structure under investigation. These results are then used in calculations based on cumulative fatigue stress hypotheses. The rainflow

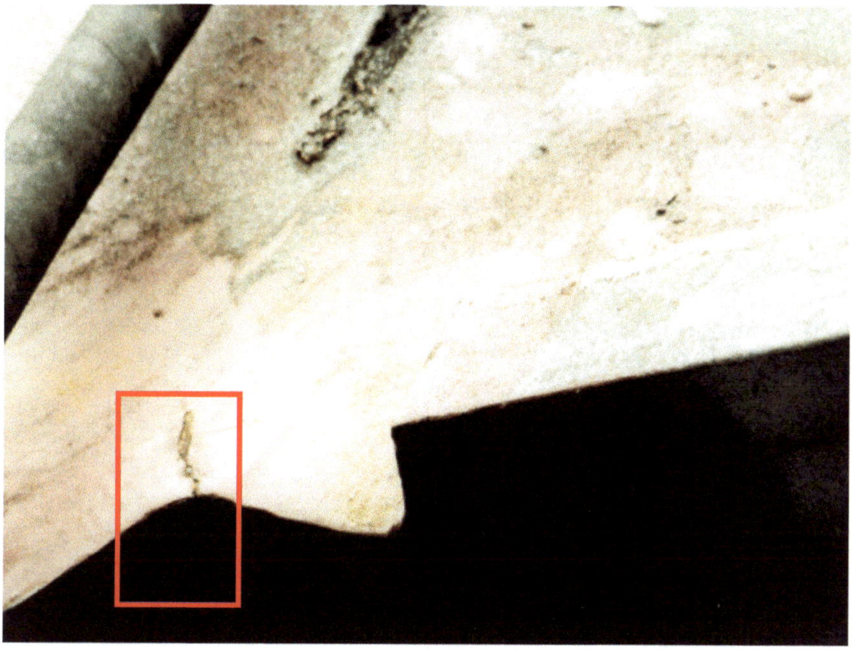

Fig. 3.20 Fracture of the structural node in the SchRs4000 bucket-wheel excavator

method is used to scale the time traces of stress changes in the machine under investigation based on the measured actual changes in loading cycle blocks. The result of calculations is the value of fatigue damage in every point of the structure per one fatigue cycle, or loading cycle block. Based on these results and on information regarding the number of hours that the machine under study has been in service, its degree of fatigue wear can be determined and thus its residual life can be estimated.

An example of this calculation procedure is presented below. The fatigue analysis pertains to a structural node in the load-carrying structure of the SchRs4000 bucket-wheel excavator, which operates in Bełchatów Coal Mine. Fatigue fractures in the parent material were detected in the shape notch of the structural joint, shown in Fig. 3.20.

These fractures occurred after 16 years of machine operation. The effective operating time of the excavator in relation to the total time in service for 16 years was 46.8%. This means that in the given period the machine worked for 65,595 operating hours. This example was used to verify the accuracy of the method of assessing the technical condition and estimating residual life of structural nodes in surface mining machines.

First, loads were identified that act on the load-carrying structure of the excavator, in particular the loads acting on the excavating unit. Dynamic loads, however, were not identified because, in the case of the excavating boom, the loads that determine the degree of fatigue wear process are generated by two forces: the circular and the lateral cutting force (F_s, F_b). The loads were measured using a scaled strain gauge

Fig. 3.21 Overburden in Bełchatów Coal Mine

monitoring system, which was installed directly on elements of the excavating unit, and measured the bending and torsion of the bucket-wheel shaft.

Data was collected while the machine was excavating different types of overburden occurring in the Bełchatów Coal Mine, which, based on a historical analysis of machine operation, were considered as representative for the almost 20-year period of machine operation. The diggability of the overburden used for tests ranged from class III to V, for which individual cutting forces range from 60 to over 120 kN/m. The view in Fig. 3.21 is an example of overburden (hard-diggable in this case), on which tests were performed in Bełchatów Coal Mine. The studies also revealed the actual location of the point of application of circular cutting force on the bucket wheel. According to the standard, the location of this force should be either in the horizontal or vertical axis of the wheel. However, studies have shown that the displacement values of the angle of application of force measured from the horizontal axis φ (Fig. 3.22), which were recorded during tests, were in the range φ = 20–30°. No correlation was observed between the value of the circular forces F_s and the angle φ. Therefore, in further calculations, the location of force F_s on the circumference of the bucket wheel was assumed to be displaced by 25° in relation to the horizontal axis of the bucket wheel (Fig. 3.22).

Based on the results of long-term analysis of loads, representative load spectra were determined for the circular cutting force, F_s, and the lateral force, F_b. Examples of time traces for both of these forces are shown in Figs. 3.23 and 3.24, respectively.

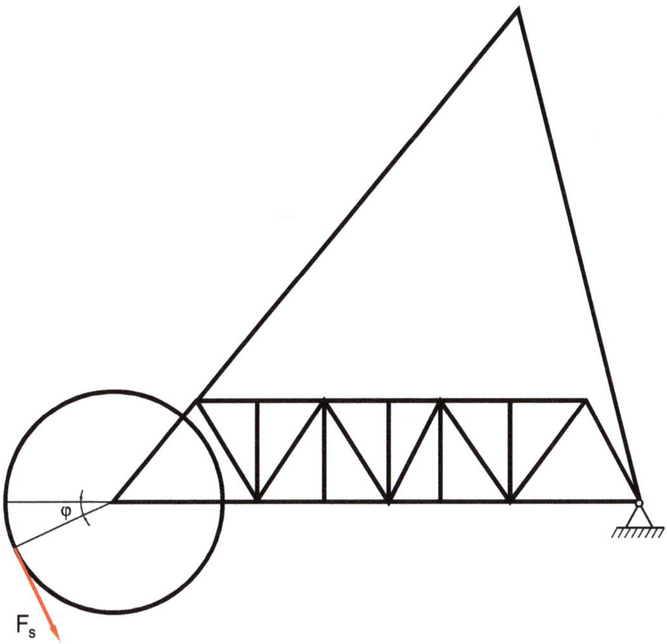

Fig. 3.22 Diagram of location of circular force F_s on the bucket wheel

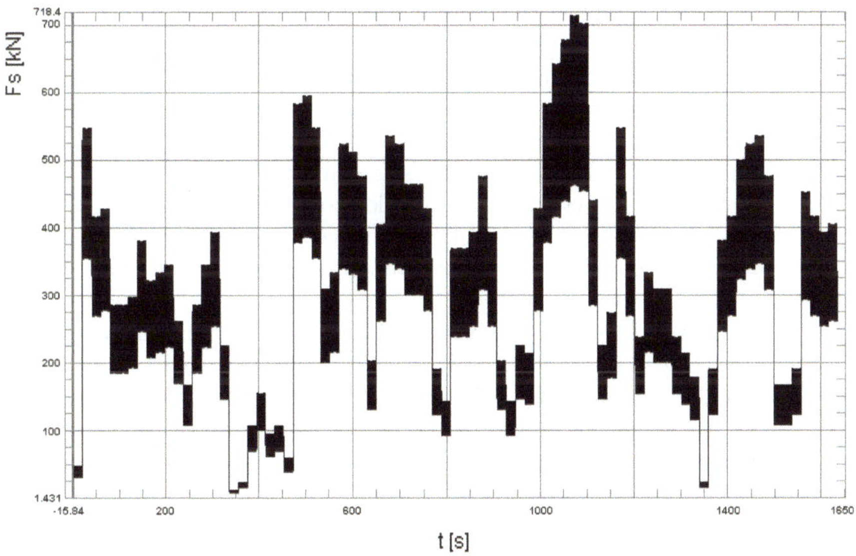

Fig. 3.23 Change of F_s cutting force over time

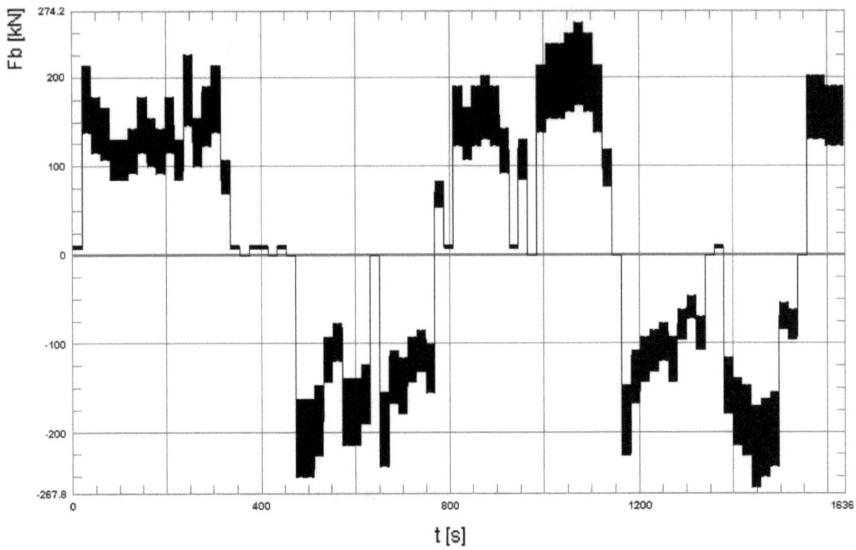

Fig. 3.24 Change of F_b lateral force over time

The next step in the developed method is the identification of the state of stress in the structural nodes under study. For this purpose, a spatial computational model of the excavating boom and the mast of the SchRs4000 excavator was created. Such a complete model accurately represents the geometric properties of the structural node as well as the complex phenomena of load transfer by subsequent elements of the excavator's spatial structure. Figure 3.25 shows the model of the excavating boom and mast of the excavator with a visible area where fatigue fractures occur.

Based on numerical calculations using the FEM, the spatial stress field was obtained for the structural node under investigation, which is shown in Fig. 3.26 in the form of contours of stress σ_x, σ_y, σ_z under load of the digging force F_s. The state of stress was also identified for the other component live loads that influence the amplitude of the fatigue cycle σ_a and for constants that influence the mean value of the fatigue cycle σ_m.

The final stage of analysis involved fatigue calculations using the cumulative fatigue damage method according to Haibach's theory. Calculations included the identified time traces of operational loads that had been recorded during experimental tests. These time traces were used to scale the state of stress identified in the numerical model for individual loads and for individual elementary cases and their fatigue combination. The calculations also included constant loads generated by the machine's mass, which influence the mean value of stresses. Therefore, an additional loading case was introduced, whose load spectrum was defined as a signal with a constant value equal to 1 and with the same period as the other live loads. Based on calculations of single fatigue damage and their accumulation, durability was determined for the node, in which fractures occur. The result of calculations is

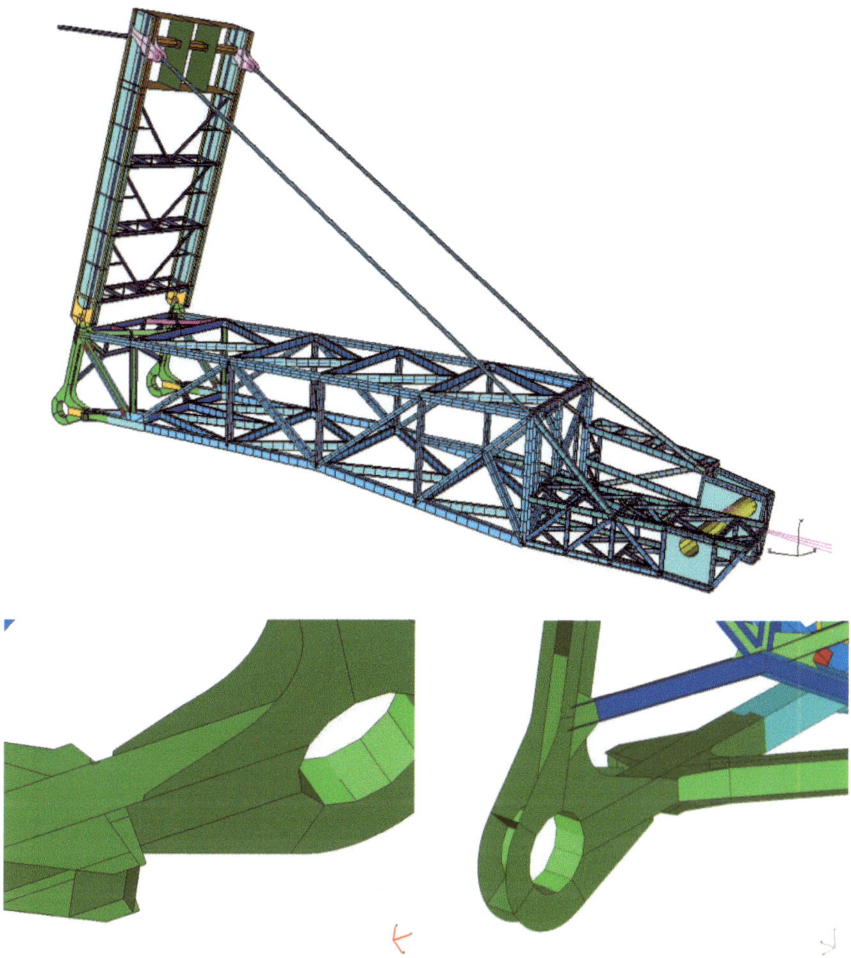

Fig. 3.25 Model of the excavating boom and mast of the SchRs4000 bucket-wheel excavator

a multiple of the spectrum of loads used in calculations until fracture occurs. In the case of a combined load spectrum and operational data pertaining to the machine's service hours, it is easier to compare the calculated operating times before fracture with the times measured on the machine in operation.

Calculations demonstrated that the area under study in the structural node operates within the range of limited fatigue strength. The calculated time of operation N_{tobl} before the occurrence of fracture was:

$$N_{\mathrm{tobl}} = 91841.3\ \mathrm{r/h}$$

Operational data demonstrated that:

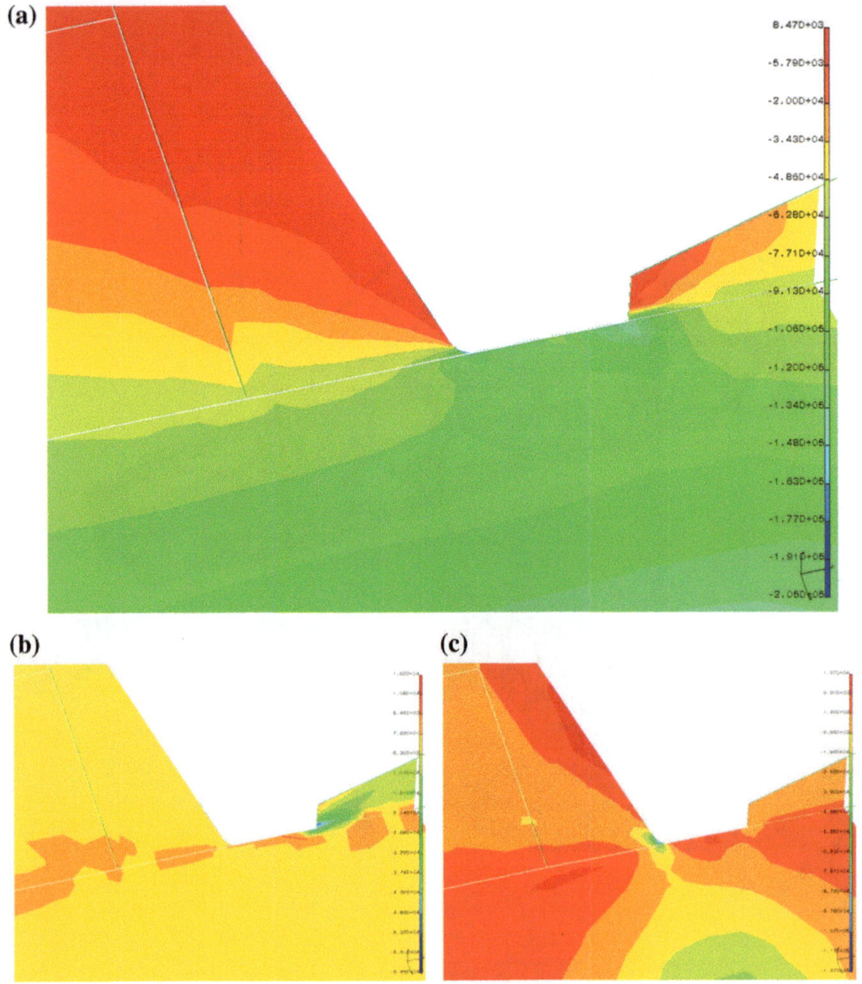

Fig. 3.26 Contours of stress σ_x (**a**), σ_y (**b**), σ_z (**c**) in Pa in the analyzed node of the excavating boom under load of force F_s

$$N_{\mathrm{t\,exp}} = 65596\ \mathrm{r/h}$$

Figure 3.27 shows contours of fatigue damage D in the analyzed area of the node in the bucket-wheel boom for a single loading cycle block. The point of origin of the fatigue fracture is clearly visible. Figure 3.28 shows a photograph of the fracture in the node and a view of the predicted location of the fatigue fracture. It can be seen that there is an excellent correlation between the model and the actual machine.

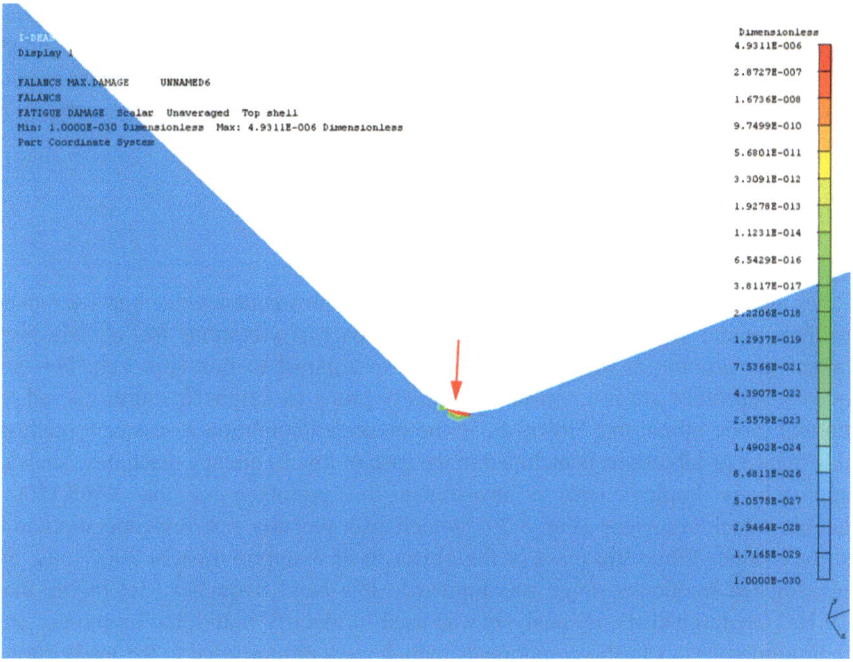

Fig. 3.27 Contours of fatigue damage D in the analyzed node of the excavator

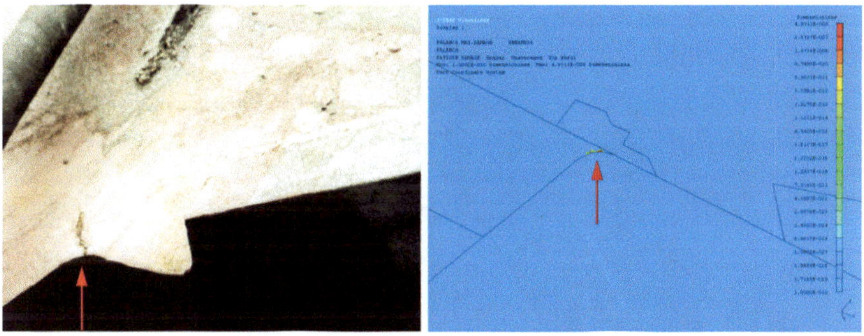

Fig. 3.28 Actual and predicted location of fatigue fracture in the analyzed node of the SchRs4000 excavator

The calculated durability is longer than the experimental lifetime. Such results were to be expected due to the fact that experimental tests on the actual machine did not include exceptional loads, which occur when the excavating unit hits non-diggable formations. Such phenomena did not occur although tests were performed on overburden with third, fourth and fifth class of diggability. Nevertheless, the results of the analysis should be considered to be satisfactory with respect to the successful prediction of the location of fatigue fracture in the load-carrying structure of the SchRs4000 bucket-wheel excavator. It should be underlined that the standard calculation procedures for estimating the durability of surface mining machines did not indicate that permissible stress will be exceeded in the node under investigation.

3.5.3 Tests on the KWK1500 Excavator

It has already been proven that the durability of high frequency machines is related to their modal characteristics and thus the impact of vibrations has already been taken into account at the designing stage. In contrast to this approach, low frequency machines are not considered equally prone to fatigue fractures caused by normal mode vibrations. However, in the presented condition assessment method, this impact of vibrations is included in the case of low frequency machines. Such an approach is implemented to investigate the prototype of the KWK1500.1 bucket-wheel excavator (Fig. 3.29), which was recently put into operation in a surface mine. While the mass of the object itself is approximately 2000 tons, the investigated frequency range was limited to low band frequency, not higher than 10 Hz. Operational modal analysis was used to identify individual harmonics and their numerous stochastic components, which were proportional to the loads during

Fig. 3.29 Object of tests—KWK1500.1 bucket-wheel excavator

excavation. However, it was also observed that the machine produces a much more uniform excitation during travel, across the whole analyzed frequency range.

Vibration is one of the most important factors affecting the service life of surface mining machines. By performing experimental and numerical tests using the FEM, it was possible to estimate the impact of normal modes of vibration on the durability of the excavator's load-carrying structure.

The uniqueness of the presented results is that for the first time in history the dynamic behavior investigated during the designing stage was compared to the dynamics of the final product. Thus far, no such analysis had been made and these complex superstructures had not previously been investigated in such detail, especially with respect to dynamic behavior.

Experimental tests

The classic method for determining dynamic properties of machines is the experimental modal analysis, which consists of exciting the structure with a forced impulse or harmonic vibration of known force and frequency [8, 9]. This can be achieved with a modal hammer or a special vibration exciter [10].

In the case of large structures, this method may be difficult to implement. Exciting vibrations in a structure such as a bucket-wheel excavator requires an enormous exciter. The method most commonly used involves the generation of an impulse by rapidly releasing a weight that is mounted on the bucket boom [8]. This method causes several problems. It requires the machine to be stopped for the time needed to prepare the testing weight and the measurement devices, which is not recommended because the nearby power plant needs a constant supply of coal (lignite). To perform the experiment, it is necessary to prepare an appropriate weight and a device that would allow the mass to be released when the desired force acts on the boom.

Operational modal analysis [11] is much simpler in use. It is used to determine the dynamic properties of the machine under investigation solely on the basis of the known system response. In order to conduct the experiment, it is therefore sufficient to place sensors and to monitor the course of acceleration during operation. There is no need to stop the machine in the preparation stage and during the experiment itself.

This form of measuring vibrations takes into account the natural operating conditions and loads. The dynamic properties of the excavator may change with the changing position of the bucket-wheel boom and body of the excavator. Operational modal analysis can be used to determine dynamic properties for different configurations of the excavator.

Monitoring system

The monitoring system consisted of accelerometers located in selected points of the machine (Fig. 3.30) and a system for acquiring and analyzing the measured data (multi-channel datalogger and analyzer). The locations of points and directions for measuring component accelerations were selected outside of mode shape nodes

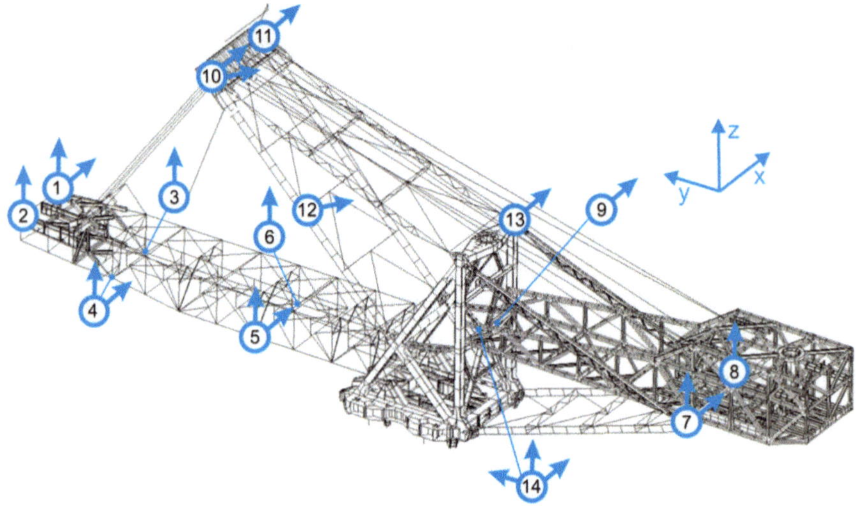

Fig. 3.30 Locations of monitoring points on the KWK1500 excavator

(unless this approach was not recommended due to the presence of a different mode shape in a given point) and also based on the authors' previous experience and research. This method of locating measurement points helps to obtain the maximum amount of information, with a minimum number of measuring devices. This is important if we consider the complicated design and large size of bucket-wheel excavators. Figure 3.31 shows an example of a measurement point with a visible accelerometer. Measurements were taken mainly during excavation. Vibration was also recorded during the machine's movements. These included machine travel with the superstructure facing the direction of travel, followed by braking to a stop, and then hoisting the bucket-wheel boom up and down during superstructure slew. Diversified forced vibration ensures that all natural mode shapes have been excited, and thus recorded.

Analyses of experimental results

The data collected during experimental tests was then analyzed in detail. First, the mode shapes and vibration frequencies were identified. In order to identify the modal model of the excavator, the Balance Realization algorithm was used [2]. The purpose of the analysis was to identify global vibrations of the entire machine and to determine the local mode shapes of its individual subassemblies.

Global vibrations were identified based on time-traces registered during digging and travel. The obtained results are presented in Table 3.2. The table also includes ratings, which help to determine the accuracy of obtained data. The Mode Phase Colinearity (MPC) criterion is the relationship between the real and imaginary parts of the determined eigenvectors. The closer the value is to 100%, the higher the probability that the identified mode shape should be classified as an Eigen-mode. In

Fig. 3.31 Example of a measuring point with accelerometers in place

practice, this criterion states that if the indicator reaches a minimum value of 90%, it can be assumed that an eigenvector is found. The Mode Phase Deviation (MPD) coefficient is the standard phase deviation in all directions of measurement. The closer the value to “0”, the higher the probability of identifying an Eigen-mode shape. In practice it is assumed that the MPD should be lower than 15°. It is equally important for the phase dispersion to be as low as possible [2]. Additionally, Fig. 3.32 shows the response spectra for selected measurement points and directions. Figure 3.33 shows the corresponding time-traces and the voltage signal of the value of load acting on the excavating unit, which is proportional to the circular digging force.

Based on data in Table 3.2, it can be concluded that the digging process excited more modes than the machine’s movements. This is due to the fact that apart from bucket-wheel rotation, the digging process also consists of travelling, hoisting and slewing. To some extent, this is a superposition of two forced vibration states. The colors in Table 3.2 mark the corresponding mode shapes identified during both measurement series. One can see changes in the frequency. The load from the conveyed material travelling across the machine causes these particular parameters to change. Other important factors include the position of the bucket-wheel boom and additional damping caused by buckets cutting overburden or lignite.

In both cases the auto-MAC coefficient was also calculated. Figures 3.34 and 3.35 show graphs of the matrix.

Table 3.2 Comparison of mode shapes of the machine for time traces recorded during digging and the machine's movements

Machine movements					Digging				
Modal shape	Frequency [Hz]	MPC [%]	MPD [°]	phase dispersion	Modal shape	Frequency [Hz]	MPC [%]	MPD [°]	phase dispersion
1	0.675	99,245	6,277	low	1	0.749	78,979	32,556	high
2	1.640	89,259	17,348	high	2	1.642	98,797	6,362	low
3	2.262	82,108	31,630	high	3	1.825	97,873	8,675	low
4	2.915	71,901	32,851	high	4	2.372	91,401	16,777	-
5	3.705	91,738	22,250	-	5	3.355	93,94	15,476	-
6	4.371	99,548	5,723	low	6	3.859	99,703	3,167	low
					7	7.022	73,32	34,516	high
					8	7.713	99,775	3,305	low
					9	8.879	80,48	35,847	high

The graphs show that the modal vectors identified during the machine's movements are much more different from each other compared to those obtained during operation. On this basis, it can be stated that the mode shapes identified in the experiment are reliable.

Figures 3.36 and 3.37 show the first two identified global mode shapes of the excavator's natural vibrations.

The first mode shape is characterized by strong longitudinal vibrations of the machine on the main mast. The amplitude of these vibrations is so large that the mode shape could often be observed with the naked eye. As shown in Table 3.1, the frequency for this mode shape is slightly different depending on the load under which it was identified.

The second mode shape consists of flexural vibrations of the bucket-wheel jib and counterweight boom in the vertical axis. The direction and phase of bucket-wheel boom vibrations are the same as for the jib. This is due to the fact that the boom is suspended by cables attached to the jib. The top view reveals a slight lateral shifting of the jib.

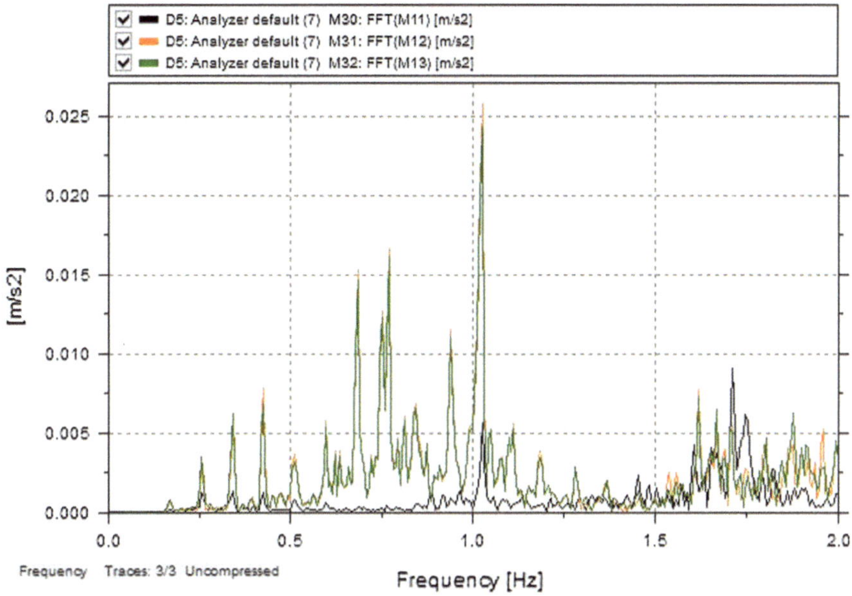

Fig. 3.32 Spectrum of vibration acceleration—point 10, 11

The next stage involved the identification of real dynamic coefficients acting on individual elements of the machine. The obtained results were evaluated based on the values of the standard dynamic load coefficient *D* [F]. It defines the static equivalent load that should be taken into account if dynamic effects are observed. The coefficient is a simple fraction, which, multiplied by the value of the static load, gives the value of the equivalent load. For easier reading and comparison, the inverse of this coefficient, 1/*D*, was used. This was calculated for individual structural elements of surface mining machines and for three mutually orthogonal directions. The standard values of the coefficient differ depending on the assumed direction and the machine element. These values are listed in Table 3.2. The higher the value in the table, the lower the equivalent load used for calculations. Based on experimental tests, the experimental value of coefficient *D* was calculated. Acceleration due to gravity, $g = 9.81\ \text{m/s}^2$, was used as static load, whereas the actual value of accelerations was calculated as the Root Mean Square (RMS) value of the vibration measured during operation. The RMS value for the digital signal is calculated using the following formula [12]

$$\text{RMS} = \sqrt{\frac{1}{N}\sum_{i=1}^{N}(x_i)^2} \tag{3.5}$$

where *N* is the number of samples. With a defined RMS value, the value of the dynamic coefficient was calculated using the following relationship:

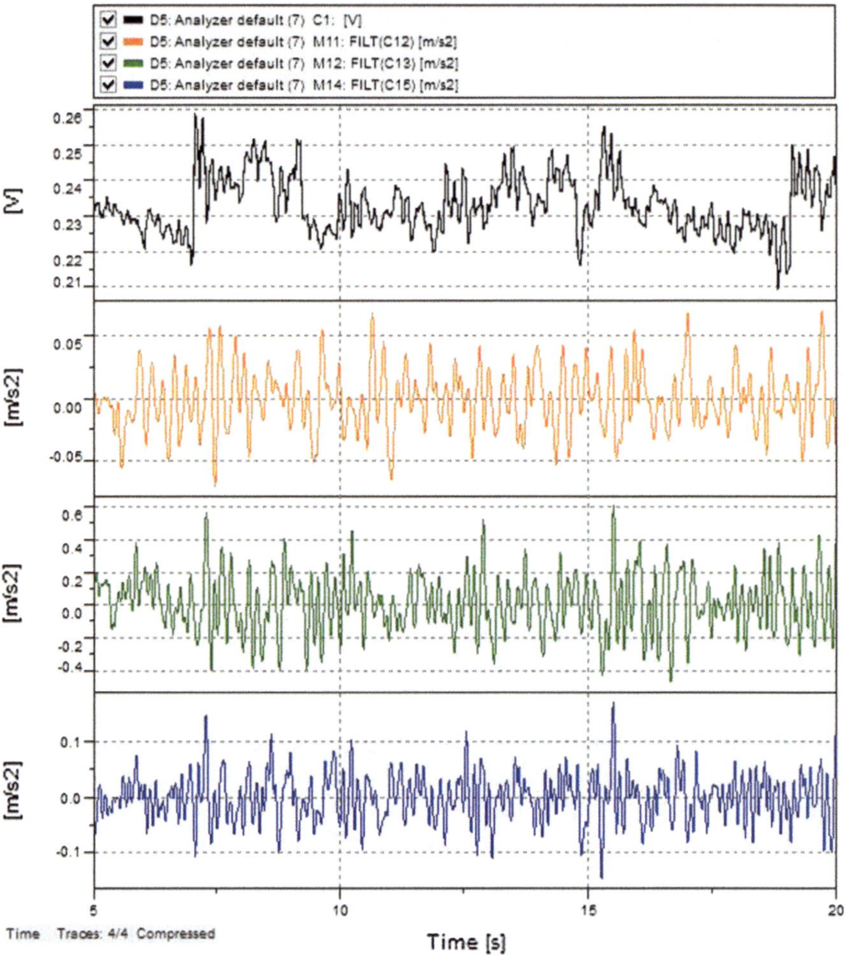

Fig. 3.33 Time-traces of vibration acceleration—point 10, 11

$$\frac{1}{D} = \frac{g}{\mathrm{RMS}} \tag{3.6}$$

The obtained data can be used to compare the standard values of the coefficient with the values calculated on the basis of measured acceleration values. The results of such a comparison are shown in Table 3.3. D_Q, D_V represent the values of the coefficient D in the lateral and vertical directions, respectively, whereas D_N is its corresponding standard value. The recommended standard values used in calculations are exceeded when the measured coefficient $1/D$ is smaller than the standard coefficient $1/D_N$. For comparison, the table also shows the values of coefficients for

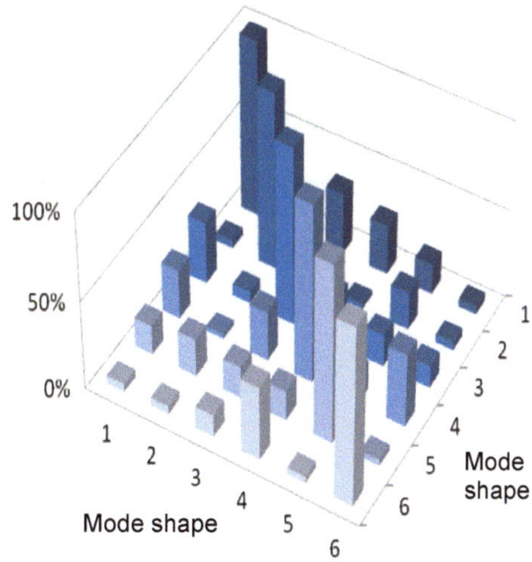

Fig. 3.34 Auto-MAC matrix for mode shapes identified during the machine's movements

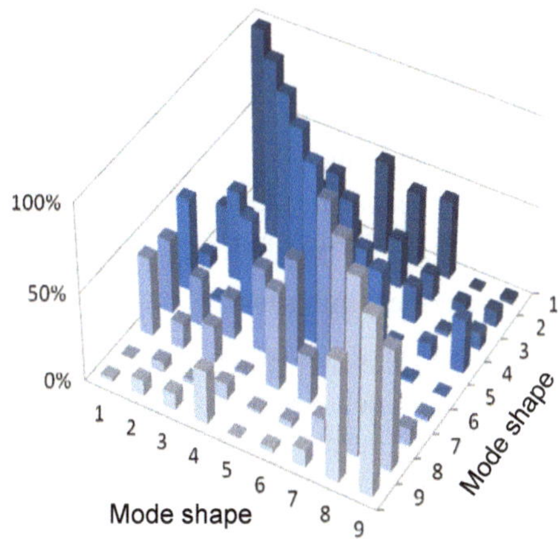

Fig. 3.35 Auto-MAC matrix for mode shapes identified during excavation

structurally similar KWK1200 and KWK1500S excavators. These tests were also performed by the authors of this book.

The comparison between standard values of dynamic coefficients and the corresponding actual experimental values clearly shows that the machine most prone to degradation, with respect to the dynamic load criterion, is the KWK1200 excavator. In two cases, this machine does not meet the required standard values. Lateral vibrations of its bucket-wheel boom and its jib exceed the permissible values. This explains the previously observed dynamic problems in the load-carrying structure of KWK1200 machines [7]. In the case of the KWK1500 excavator, the standard criterion has not been met in only one case: lateral vibrations of the bucket-wheel jib. Particular attention should be given to the fact that the same criterion is not satisfied for both excavators. Additionally, it is evident that the ratios between the standard value of vibration in the bucket-wheel boom and the measured values are

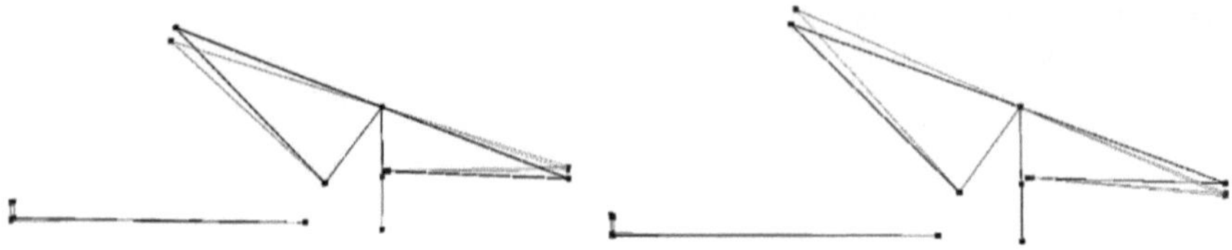

Fig. 3.36 First mode shape of vibration

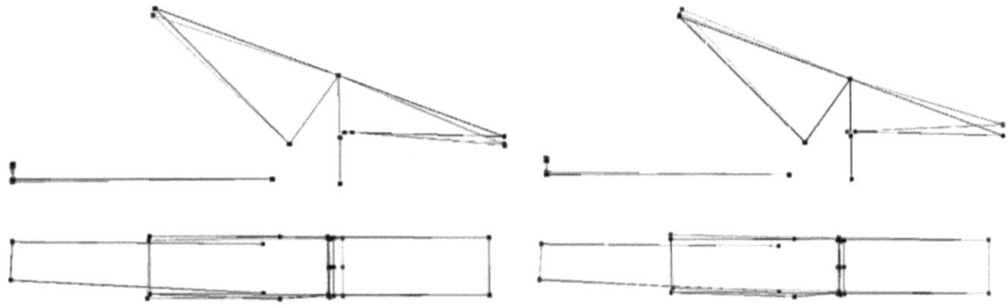

Fig. 3.37 Second mode shape of vibration

very similar. In the case of the KWK1500.1 excavator, such a phenomenon was not observed, despite the almost identical shape of its superstructure in comparison with the KWK1500 excavator.

Numerical examinations

Numerical examinations involved the construction of a calculation model of the excavator, which was used to determine the correctness of obtained numerical and experimental results and to evaluate the influence of dynamic loads on the durability of the load-carrying structures of machines. Figure 3.38 shows this model. It is a shell and beam model.

In order to evaluate the influence of dynamic loads on the durability of the load carrying structure in the machine under investigation, fatigue damage calculations were performed for the KWK1500.1 excavator superstructure using the Finite Element Method and numerical models of individual elements in the machine.

Calculations were performed using the actual experimental values of dynamic coefficients, which occur during operation. These values were applied in the numerical models. Fatigue calculations were carried out in accordance with the design standard [13]. Table 3.4 presents examples of results of such calculations for the counterweight boom, in which the influence of dynamic loads on durability was the largest. For comparison, the table includes values of the total stress range for the fatigue case and analogous values for dynamic loading cases D_q and D_v and their percentage share in the combined loads. Figures 3.39 and 3.40 also compare the contours of the fatigue stress range.

Table 3.3 Comparison of coefficients of dynamic loads

Machine element	Type of D coefficient	Experimental value $1/D$			Standard value $1/D_N$	D_N/D		
		KWK1200 [14]	KWK1500S [15]	KWK1500.1		KWK1200	KWK1500	KWK1500.1
Bucket-wheel boom	Average D_Q	45	97	93	60	0.75	1.62	1.54
	Average D_V	22	41	20	10	2.20	4.10	2.00
Bucket-wheel jib	Average D_Q	37	45	208	60	0.62	0.75	3.46
	Average D_V	35	53	32	10	3.48	5.30	3.18
Counterweight boom	Average D_Q	132	88	168	30	4.40	2.95	5.61
	Average D_V	71	84	30	25	2.84	3.37	1.22

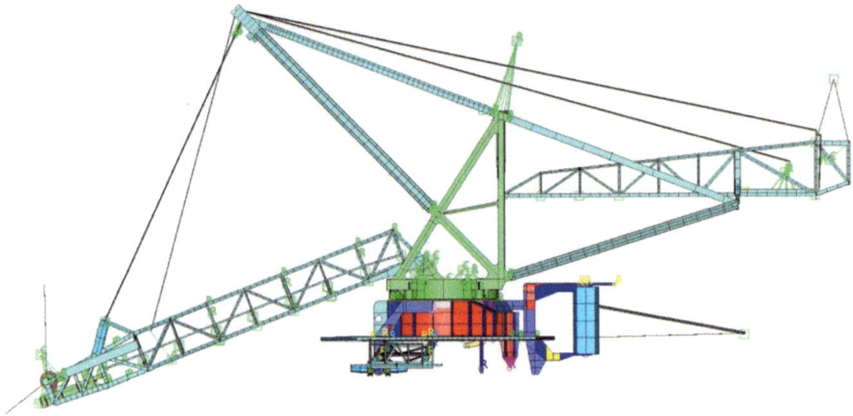

Fig. 3.38 Numerical model of the KWK1500.1 excavator

Table 3.4 Stress ratio of dynamic loads D_q and D_v for standard fatigue load combinations

Location (see Figs. 3.39 and 3.40)	H1b—stress range (combined fatigue loads)	D_q		D_v	
	[MPa]	[MPa]	H1b [%]	[MPa]	H1b [%]
1	139	24	17	19	79
2	24				

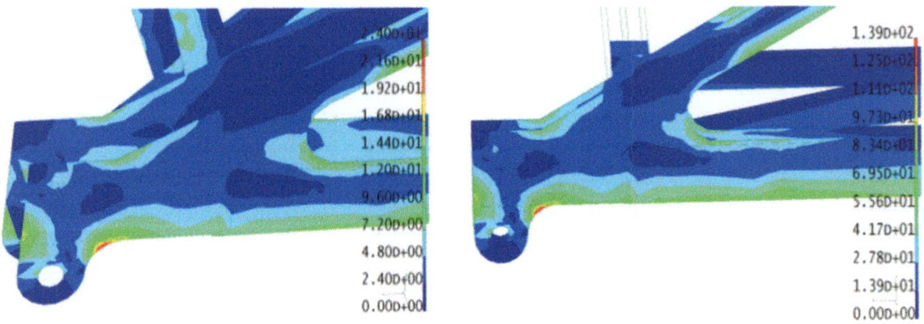

Fig. 3.39 Ratio of dynamic loads D_q (*left*) in total fatigue stress (*right*) [MPa]—area 1 (see Table 3.4)

Based on the analysis of results from experimental research, the actual values of dynamic coefficients were established for individual elements of the machine, which were compared to the standard values used by designers. In order to evaluate the influence of dynamic loads on the durability of the load-carrying structure in the excavator under investigation, numerical calculations were performed using the FEM. The results of examinations led to the following conclusions:

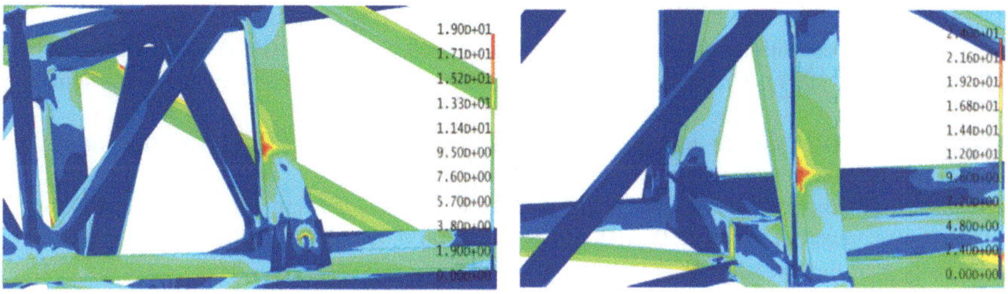

Fig. 3.40 Ratio of dynamic loads D_v (*left*) in total fatigue stress (*right*) [MPa]—area 2 (see Table 3.4)

1. In some cases, the standard values of dynamic coefficients are lower than the actual dynamic loads.
2. The method used by the authors to construct numerical models of surface mining machines and to assess their condition, helps to obtain a description of dynamic phenomena that corresponds to real machines.
3. The influence of dynamic loads on the durability of the load-carrying structure is diversified for individual elements of surface mining machines.
4. The largest proportion of dynamic loads, reaching 80%, was found in the structure of the counterweight boom, for loads acting in the vertical direction.
5. It was confirmed that low-frequency vibrations are the main cause of potential fatigue damage to components of the elements in the load-carrying structure of a bucket-wheel excavator.

References

1. DIN 22261-2 Excavators, spreaders and auxiliary equipment in opencast lignite mines—Part 2: calculation principles
2. Uhl T, Lisowski W, Kurowski P (2001) In-operational modal analysis and its applications. AGH Publishing House, Kraków
3. Alenowicz J (2004) Diagnostics of the loads carrying structures of surface mining machinery. Górnictwo Odkrywkowe. No 2
4. Dudek D (2004) Nature of the degradation of the technical objects. Problemy Rozwoju Maszyn Roboczych
5. Dudek D, Hawrylak H (1995) Degradation process of the earth moving machinery after long operation. Maszyny-Dźwignice-Transport. No 3
6. Aicher WJ (1976) Description of a fighter aircraft loading standard for fatigue evaluation "Falstaff". Common report of F+W Emmen, LBF, NLR, IABG
7. Czmochowski J (2008) Identification of modal models of excavators in lignite mining. Wroclaw University of Technology Publishing House
8. Arczewski K, Pietrucha J, Szuster J (2008) Vibrations of physical systems. Warsaw University of Technology Publishing House, Warszawa
9. Piszczek K, Walczak J (1982) Vibrations in machine design. PWN, Warszawa
10. Uhl T (1997) Computer aides identification of models of mechanical structures. WNT, Warszawa

11. Hermans L, Van der Auweraer H (1999) Modal testing and analysis of structures under operational conditions: industrial applications. Mech Syst Signal Process 13(2):193–216
12. Lyons RG (1999). Understanding digital signal processing. Transport and Communication Publishers
13. PN-G-47000-2 Surface mining. Bucket wheel excavators, vol 2 (Calculation principles). Polish Standardization Council
14. Technical Report IMDO. PWr. 2009, Ser. SPR ; no 91; Testing and diagnostics of bucket wheel excavators KWK-1500S (K-9), KWK-1200M (K-14), SCHRS-1200 (K-22), vol. 2. Testing of the excavator KWK-1200 (K-14)
15. Technical Report IMDO. PWr. 2009, Ser. SPR ; no 92; Testing and diagnostics of bucket wheel excavators KWK-1500S (K-9), KWK-1200M (K-14), SCHRS-1200 (K-22), vol. 2. Testing of the excavator KWK-1500 (K-9)

Chapter 4
Failure Analysis

During long-term operation, surface mining machines [1] are susceptible to minor or major failures resulting from fractures in the load-carrying structure or in mechanical elements [2–4]. Such fractures can be caused by ultimate or fatigue stress. Fatigue fractures should be identified beforehand by means of methods described in the previous chapter. Fractures can be divided into those that do not affect the load-carrying capacity or stability, and those whose growth is a significant threat to the load-carrying capacity. Whether a fracture is a threat to the load-carrying structure is determined by evaluating the stress state of a given area and the role of a given element. In a properly designed load-carrying structure, each element performs a specific function. Moreover, there are fractures that do not grow during further operation (e.g., due to relaxation of local stresses) or that stop growing due to special treatment; for example, the drilling of holes. The reason that a fracture does not lead to serious consequences is because the load-carrying structure is dimensioned with a safety margin. Although the methods of analyzing stress state finite element method (FEM) are increasingly more accurate, it is still not possible to design a load-carrying structure without locally increasing stresses.

The occasional failures of these machines are most frequently caused by the following:

- Operational faults of machine operators, e.g. cutting too deep into the slope (Fig. 4.1), excavating excessively thick sections of earth, driving the bucket-wheel boom into the slope, chute blockage, using machine mechanisms for purposes other than intended.
- Lack of overload protection systems or their improper setup, e.g., bucket-wheel drive mechanism, superstructure swing mechanism, travel mechanism, and discharge control.
- Structural faults (structural notches (Fig. 4.2), excessive concentration of welds (Fig. 4.3), underestimation of actual loads, failure to take all loads into account, and sections that are too small).

E. Rusiński et al., *Surface Mining Machines*,
DOI 10.1007/978-3-319-47792-3_4

Fig. 4.1 Buckling of bucket-wheel boom frame due to slope landslide [2]

Fig. 4.2 Shearing of the crawler drive shaft in the place of stress concentration

- Assembly faults, e.g., forced assembly with preliminary stress, lack of assembly supports, risk of loss of stability, and failure to comply with assembly technology.

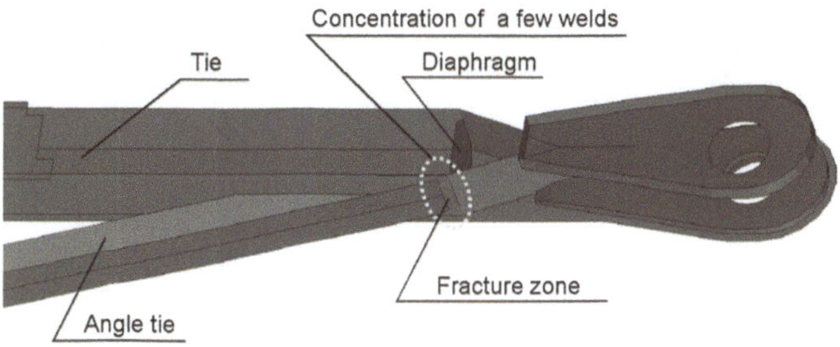

Fig. 4.3 Excessive concentration of welds in the counterweight boom support tie member of the KWK-1400 excavator

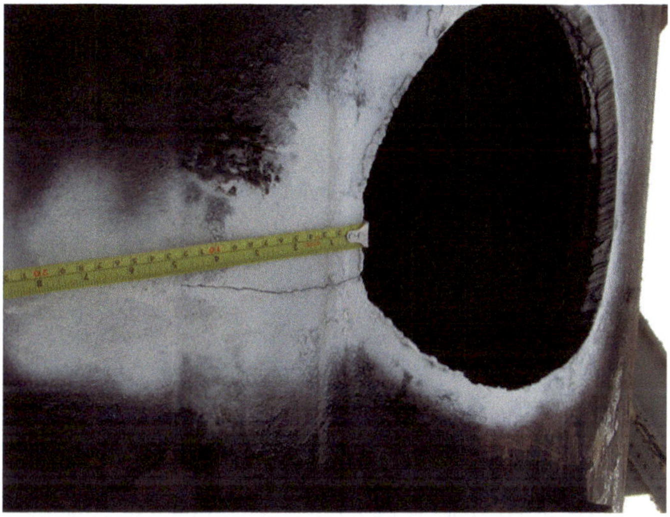

Fig. 4.4 Incorrect cut in the load-carrying column of an Rs-560 bucket-chain excavator, resulting in a fracture over 10 cm in length

- Technological faults, such as incorrectly welded joints, incorrectly made or located cuts (Fig. 4.4), e.g., for electric wiring or lubrication units.
- Resonance vibrations [5] generated by the excavation process, when natural vibrations of the superstructure overlap with vibrations generated by each bucket cutting into the soil (i.e., the excavated material) [6–11].
- Incorrect technology for repairing damaged or worn elements, e.g., regeneration of shafts using the hard-facing technique for worn slide surfaces (Fig. 4.5).
- Machines catching fire, e.g., conveyor belts catching fire due to friction on worn idler.
- Geological, e.g., slope and bench slides with machines on them.

Fig. 4.5 Fatigue fracture of an axle in the crawler system of A2RsB-12500 stacker regenerated using the hard-facing technique

- Absence of, or improperly performed, periodic technical inspections, or inspections of the location of permanent damage symptoms (e.g., fractures, plastic deformations, torn rivets or bolts).

4.1 Methods

Failures of surface mining machines lead to substantial financial losses [12] related to the destruction of the machine, and losses resulting from the interruption of the excavation plan. Thus, the user (owner) tries to restore the state of the machine's systems as quickly as possible. A failure is followed by tests aimed at identifying the cause of the failure, which include the following: assessment of the condition of the machine after the failure, computer simulations of the effort and failure, and material studies [13, 14].

In order to identify the causes of a failure. it is necessary to assess the state of the machine's systems after the failure. The first stage of this process involves an on-site visual inspection and interviews in order to determine the circumstances of the failure, which comprises taking records of the following:

- Photographic documentation (detailed and general),
- Geometry measurements (e.g., global and local, measurements of thickness and diameters),
- Reconstruction of the position of the bucket-wheel boom (rotation and lift) and the discharge boom or bridge,
- Measurements of the pit, slope state and soil stability,
- Type of excavated material (e.g., overburden, coal, rock inclusions),
- Atmospheric conditions during failure (e.g., temperature, wind speed),

- Analysis of periodic inspections,
- Age of machine,
- Analysis of renovations, repairs, maintenance activities.

The machines currently operating in lignite coal mines are highly worn and prone to fatigue fractures or subjected to excessive overloads in order to increase productivity, among other things. Using calculation possibilities of contemporary programs and computers, it is possible to accurately determine the stress state of such complicated systems in these machines. In order to understand the causes of failures, strength calculations of the load-carrying structure are usually performed using the FEM. To reliably recreate the effort of the superstructure, it is necessary to identify the loads that can occur during failure [9, 15–18]. Additional calculations are performed in accordance with the current standards [19]. The accuracy of current calculation models makes it possible to determine not only the stress state of main girders, but also the stress state of structure nodes, i.e. places where these girders are connected. Due to the fact that such structures—in particular welded structures—are prone to fatigue cracks, apart from the stress state, calculations also include stress changes. An example of a calculation model of a tie member that broke during the catastrophic failure of the KWK-1400 excavator is shown in Fig. 4.6, and the detailed view of the fractured area is shown in Fig. 4.7. Figure 4.8 presents the distribution of changes in the stress range for the H1b case [19], which

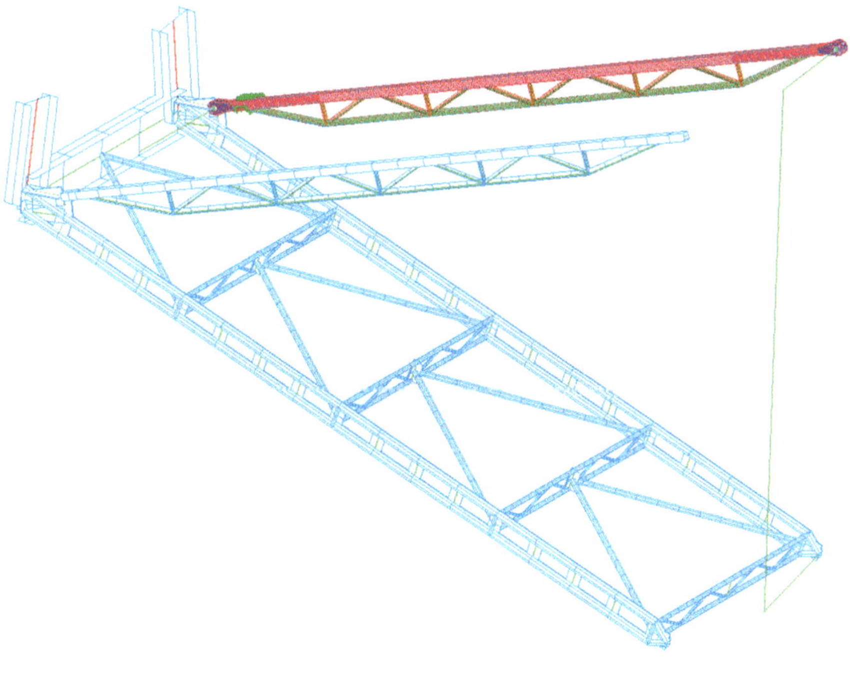

Fig. 4.6 Discreet model of counterweight boom jib in a bucket-wheel excavator

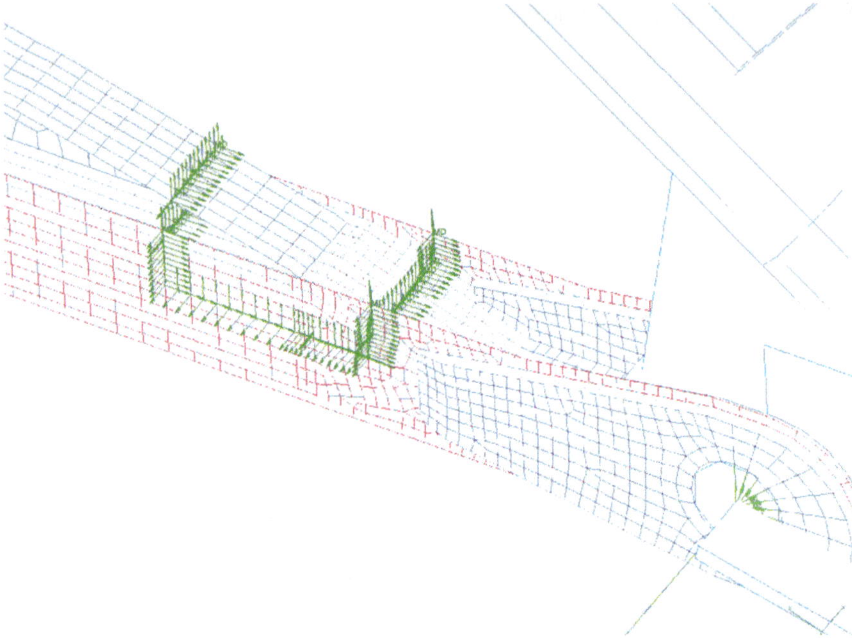

Fig. 4.7 Fragment of a discreet model of the jib in the vicinity of the fractured zone

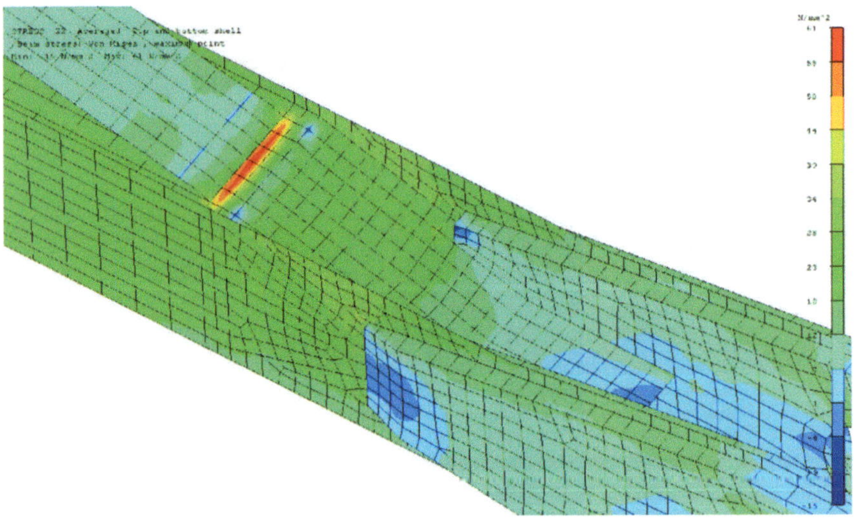

Fig. 4.8 Contours of nominal stress range Δσ in the *X* direction (along the direction of tie member operation) on a tie member fragment in the H1b case

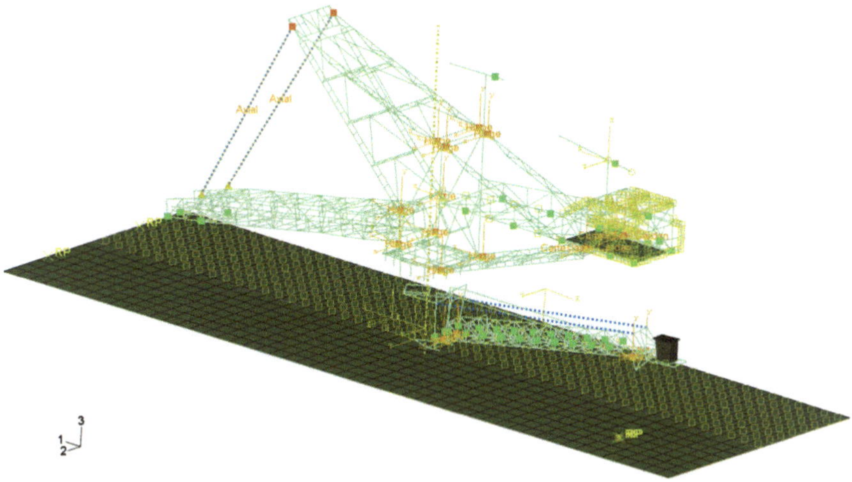

Fig. 4.9 Model of the load-carrying structure of the KWK-1400 excavator for simulation of failure

is a meaningful indicator when assessing fatigue strength. The presented stress distribution clearly shows that in the area of the fracture there is a concentration of stress, which occurs in an unfavorable location, i.e. on the welded joints. Therefore, the concentration of stress occurs in areas whose structure may have been altered by welding, which renders them prone to fatigue cracks.

Computer methods can also be used to simulate the course of the failure. This is crucial when evaluating the consequences of the failure and rebuilding the load-carrying structures of the machine using seemingly undamaged elements. Figure 4.9 depicts the calculation model of the KWK-1400 excavator load-carrying structure used for simulation of the failure. When creating such a model and simulation, it is important to recreate the configuration of individual assemblies in the moment of failure, i.e. the position of the bucket-wheel boom, the position of the superstructure in relation to the undercarriage and the mutual positioning of the discharge boom relative to the counterweight boom. The problem under consideration in this case is a dynamic and time-varying issue, which occurs due to the fracture of the tie member that supports the counterweight boom. A computer simulation makes it possible to observe the behavior of the load-carrying structure during subsequent time intervals after the fracture of the tie member, i.e. to monitor the state of deformations and stresses in particular. Figure 4.10 shows the stress state of the structure whereas Fig. 4.11 shows the state of plastic deformations after the failure. The identification of the causes of a failure may serve as the basis for classifying a given element for use in the reconstruction of the machine's system. Permanently deformed elements are excluded from the repair process.

Material tests are crucial in assessing the direct causes of machine failures. Basic methods of analyzing elements in the fracture point include the following:

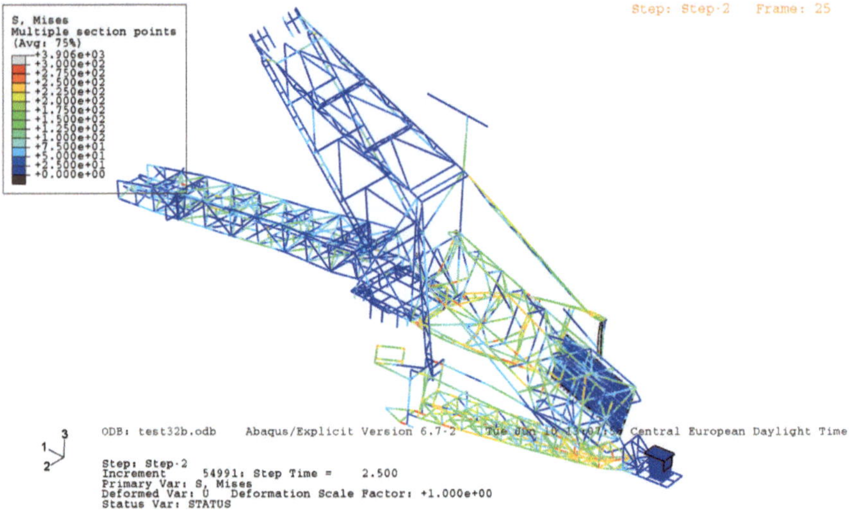

Fig. 4.10 Equivalent stress according to Huber-Mises hypothesis (MPa) 2.5 s after the fracture of the tie member

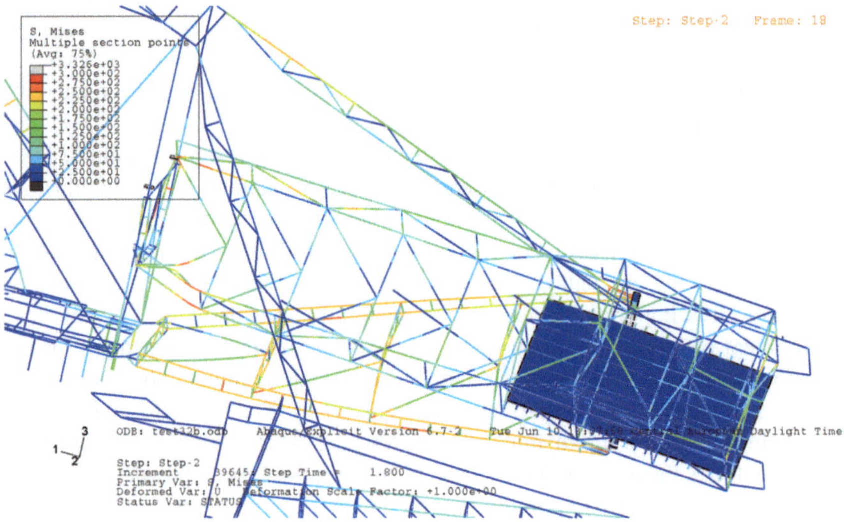

Fig. 4.11 Plastic deformations in the structure, elapsed time: 1.8 s

- Macroscopic examination (analysis of fracture),
- Microscopic examination (structure in the vicinity of the fracture),
- Microindentation tests,
- Analysis of chemical composition,
- Analysis of mechanical properties.

Based on macroscopic examinations, it is possible to establish if the fracture is the result of ultimate or fatigue stress. An analysis of the microstructure helps to determine if the microstructure is appropriate for a given grade of steel in relation to the recommended plastic or thermal processing. The results of such analysis can be used to identify structures prone to fatigue cracks; for example. in the heat affected zone (HAZ) near the welded joints. Microscopic examinations are complemented by microhardness measurements, which help to identify martensite areas characteristic of overheated areas in the HAZ. The purpose of chemical composition analysis and tests of mechanical properties is to verify if the steel grade selected for the load-carrying structure was appropriate and had appropriate strength properties.

The complete set of data obtained through the above examinations helps to identify the cause, the course and consequences of a failure. Such data are sufficient to choose the scope and method of renovating the damaged machine. In some cases, it might be decided to scrap the entire machine or its parts.

The following section describes examples of research into failures of surface mining machines.

4.2 Examples

4.2.1 KWK 1400 Excavator Collapse

An example of a severe failure of the counterweight is the KWK 1400 excavator collapse, which occurred after the fracture of the counterweight boom support tie member. While the excavator was digging overburden and the bucket-wheel boom was in its low position, the left tie member in the counterweight boom support broke (Figs. 4.12 and 4.13). In the analyzed case, a numerical model of the whole excavator was created in order to simulate the collapse, and detailed models of the counterweight boom support were created in order to analyze the state of stress. This was accompanied by tests of material in the fracture zone in order to investigate the causes of the fracture.

One of the main objectives of FEM calculations in this case was to identify the spatial stress field generated by operational loads of the excavator and to determine its impact on the durability of tie members. Most of the structural elements were modeled using beam elements. One of the two tie members of the jib was modeled using shell elements (Figs. 4.14 and 4.15).

Stress calculations were carried out for loads defined in standard [19]. Static and alternating loads were analyzed, which the standard defines as H1a and H1b, respectively. The sources of alternating loads include dynamic forces, mass forces generated by technological movements, tilting, loads generated by excavated material, digging forces, and so on [20]. The tie member has a box structure with an additional lower truss whose function is to stiffen the tie member that counteracts

(a)

(b)

Fig. 4.12 KWK-1400 excavator: **a** before the collapse, **b** after the collapse

lateral vibrations [7]. The drawback of this solution is that the neutral axis is shifted towards the truss, which results in the eccentric action of stretching forces. This eccentricity of forces bends the tie member upwards in the vertical plane, as shown in Fig. 4.16.

The results of strength calculations confirmed that stress concentration occurs in the fracture zone for the H1a case (Fig. 4.17) and that the stress generated by H1b

Fig. 4.13 Fracture point of counterweight boom support tie member

Fig. 4.14 Placement of the fracture

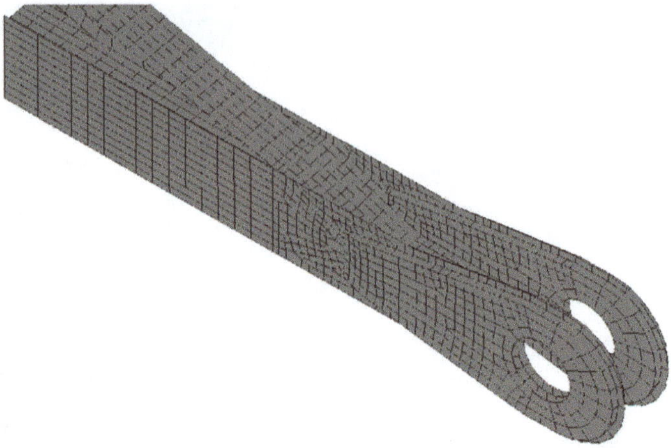

Fig. 4.15 Discrete model of the node where the fracture occurred

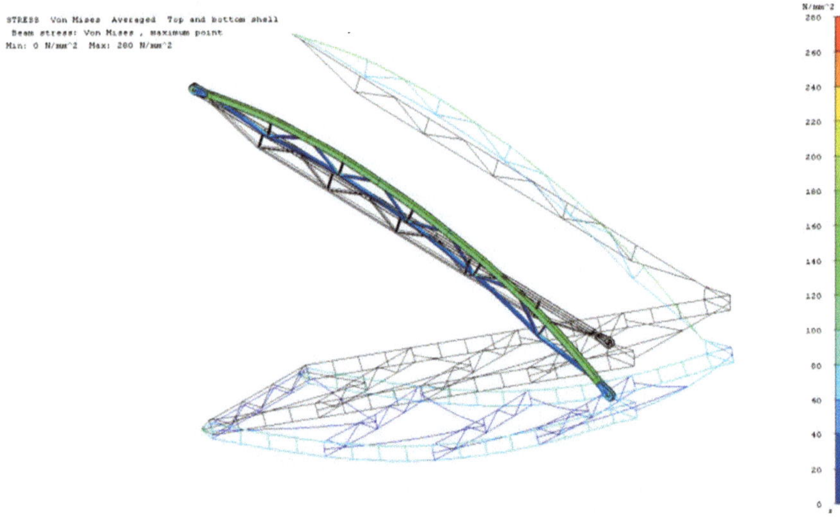

Fig. 4.16 Contours of average equivalent stress in the jib model for the H1b case

alternating loads exceeds permissible values, whose definition in the standard is based on the notch class of the welded joint, the level of average stress and the range of stress change. The average stress in the point of concentration was $\sigma_m = 302$ MPa for the amplitude of stress changes $\sigma_a = 31$ MPa. Permissible stress for this case is $\sigma_{dop} = 315$ MPa, which means that it is exceeded in cycles of changes from σ_{min} to σ_{max}, since $\sigma_{max} = \sigma_m + \sigma_a = 333$ MPa.

Additional calculations involved the simulation of a fracture that progressed through 50% of the tie member's cross-section. The results are shown in Fig. 4.18.

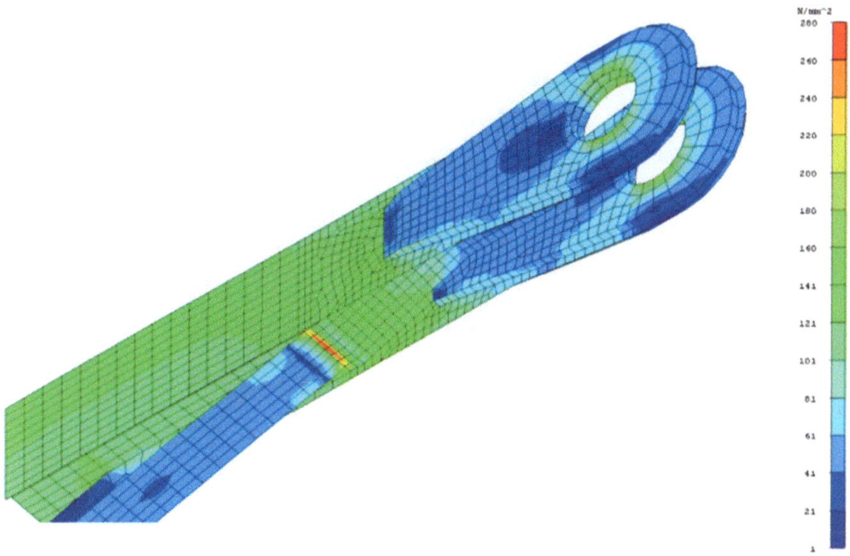

Fig. 4.17 State of stress in the tie member in the vicinity of the fracture

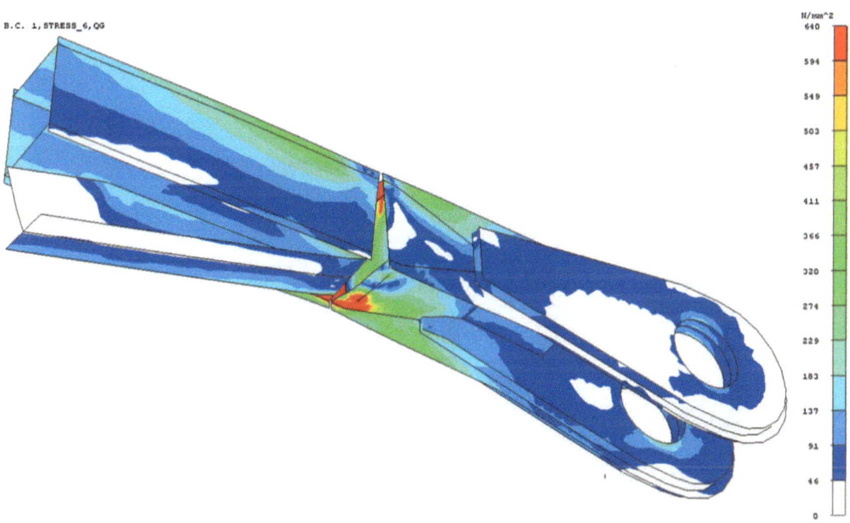

Fig. 4.18 Contours of HMH equivalent stress on the model with the crack for the H1a load case

It was found that such reduction of the tie member's cross section would lead to its final overload fracture (shown in Figs. 4.19 and 4.20). This, in fact, did result in the collapse of the excavator (Fig. 4.21).

In light of the calculation results, the most probable cause of damage to the tie member was the fact that permissible stress was exceeded in the vicinity of the weld

Fig. 4.19 View of the damaged cross section of the tie member on the mast

Fig. 4.20 View of the fractured tie member of the excavator on the counterweight eye

Fig. 4.21 General view of the excavator collapse and the damaged counterweight

connecting the bottom flange of the lower truss girder with the tie member wall. The stress concentration was a result of structural factors, i.e. the local bending of the tie wall in the vicinity of the weld caused by the action of the tie truss shelf, the diaphragm and the bending of the wall near the diaphragm (Fig. 4.22), as well as the notch introduced by the weld and dependent on the weld class. It should be mentioned that the values obtained analytically (e.g., using an analytical model of the tie member in the form of a truss) would not take into account the stress field that was found using the presented FEM calculations. Such calculation methods were not available when the machine was being constructed (in the early 1980s) and it was impossible to verify if this welded connection, which resulted in tie member fracture, was suitable.

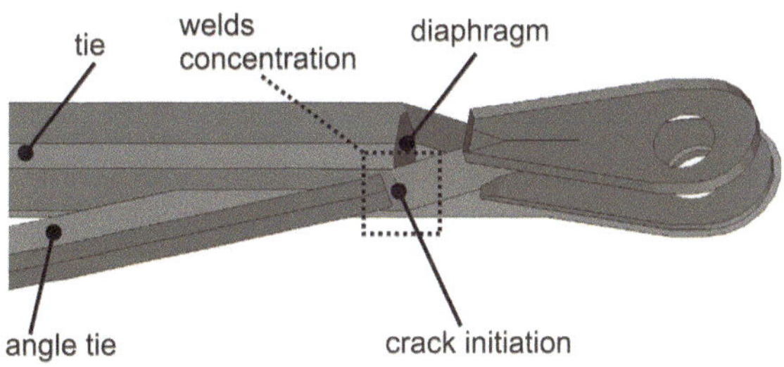

Fig. 4.22 Fracture point of counterweight boom support tie member

Samples were taken from the fracture zone to perform macroscopic and microscopic examinations (Fig. 4.23). Numbers 1, 2, 3, and 4 mark the plates chosen for examinations.

The macroscopic image shows fatigue lines and delamination of material in the brittle fracture zone. The fatigue zone covered the whole cross section of plates 1 and 2. The brittle fracture zone was present in plates 3 and 4. The fatigues lines propagating in the fatigue zone of plates 1 and 2 indicate that the fracture originated from the vicinity of the welded joint between plates 1 and 2 (site A in Fig. 4.24). An enlarged image of site A is shown in Fig. 4.25a. The arrow marks the probable

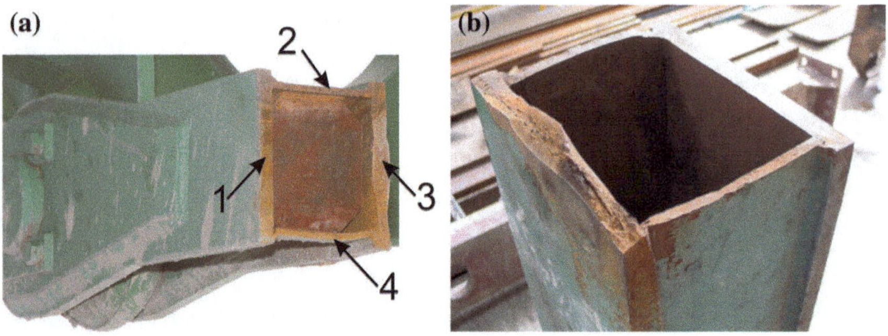

Fig. 4.23 Samples of the tie member supplied for examinations: **a** on the pin, **b** on the tie member with truss

Fig. 4.24 View of the fracture in the tie member. *A* and *B* marks the areas chosen for further macroscopic and microscopic examination

initiation site of the fatigue crack, from which the fracture propagated. This site was subject to direct interaction between the fillet weld connecting plates 1 and 2 and the welds connecting plate 2 with the truss bracing. Figure 4.25b shows an enlarged fragment of the fracture in site B with the marked fatigue and brittle zones and with visible delamination of material in the brittle zone. The morphology of the fatigue zone of the fracture in site A on plate 1 observed through a scanning microscope is shown in Fig. 4.26a. The morphology of the brittle zone in site B on plate 3 with visible plastic deformations and delamination of material is shown in Fig. 4.26b.

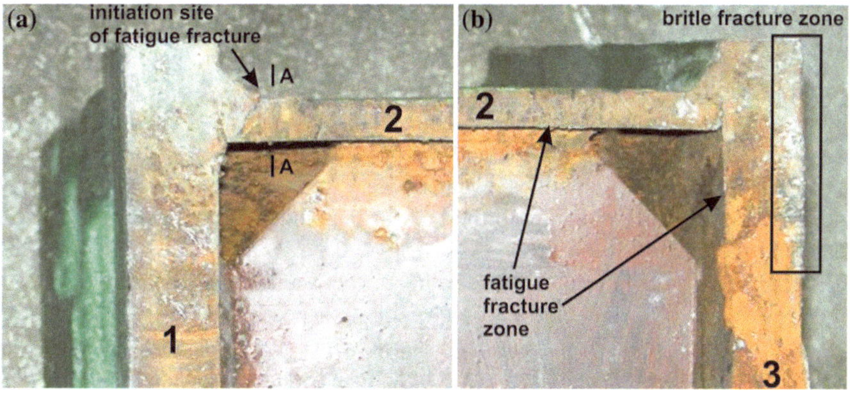

Fig. 4.25 Enlarged fragment of the fracture in sites *A* and *B*

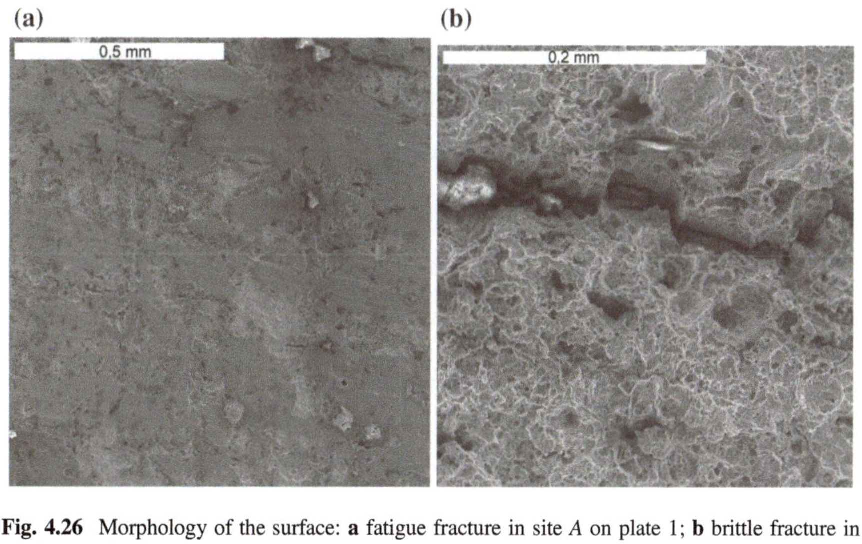

Fig. 4.26 Morphology of the surface: **a** fatigue fracture in site *A* on plate 1; **b** brittle fracture in brittle zone in site *B* on plate 3. SEM

The microstructure of the joint between plates 1 and 2 was examined in detail in areas marked in Fig. 4.27. The material of plate 1 had a ferritic and pearlitic structure (Fig. 4.28a), but the microstructure in the cross section was highly non-homogenous, which is shown in Figs. 4.28b and 4.28c.

The microstructure of area 2, 3 and 4 is presented in Fig. 4.29.

In the vicinity of the welded joints, a ferritic-pearlitic structure was identified with characteristics of the Widmanstätten pattern (Fig. 4.31), i.e. brittle areas, which were the initiation site of the fatigue crack. These bands could be the result of welding faults; for example, from cooling that was too rapid.

Microscopic examinations were performed for the A-A cross section marked in Fig. 4.25a, which runs near the initiation site of the fatigue fracture. The plane of the micro-section was perpendicular to the surface of the fracture and to the weld connecting plate 2 with the bracing. The micro-section also ran longitudinally to the weld between plates 1 and 2. In the area of the micro-section, there were impacts from the welds between plates 1 and 2, and between plates 2 and the bracing or interaction between two HAZ. The fracture in plate 2 probably progressed into the HAZ of the weld between plate 2 and the bracing.

The microscopic examinations revealed a very narrow area, which could indicate the presence of HAZ belonging to the end (or beginning) of the weld between the plate and the bracing.

The HAZ was partly located in the weld area between plates 1 and 2. This area is indicated by an arrow in Fig. 4.30 and exhibits a pearlitic-ferritic structure with characteristics of the Widmanstätten pattern, which confirms that the area was melted or excessively heated. In the studied area, "islands" of darker etching were observed, which might indicate the presence of martensite or bainite. In order to identify martensite areas, Vickers microindentation tests were performed under a

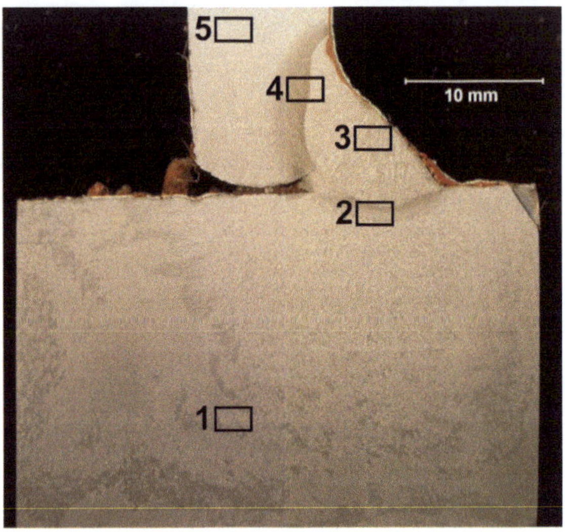

Fig. 4.27 Microstructure of the welded joint with places examined in detailed. Etched with Mi1Fe, optical microscope

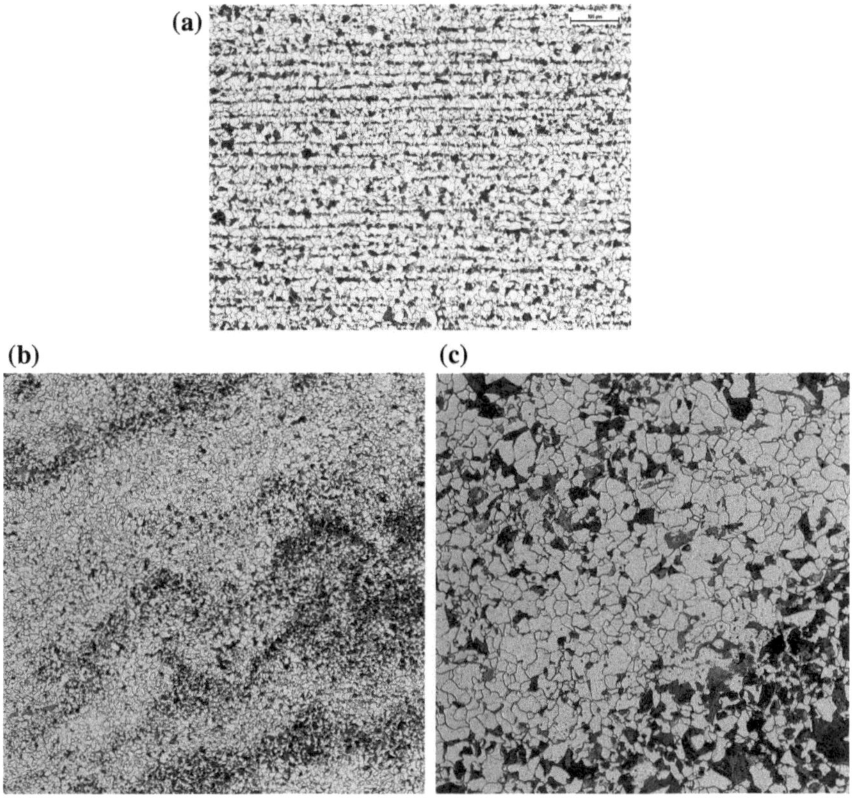

Fig. 4.28 Microstructure of plate 1 in area 1 shown in Fig. 4.24. Etched with Mi1Fe, optical microscope: **a** ferritic-pearlitic structure with visible banding; **b** and **c** micro-section performed perpendicularly to the cross section shown in Fig. 4.27, longitudinally and laterally to the axis of the tie member and plate. Ferritic-pearlitic structure with visible non-homogeneity

load of 50 g. Figure 4.31 shows the enlarged view of the microstructure in the area indicated by an arrow shown in Fig. 4.30.

The tests in the discussed area showed that the permissible hardness of 350HV was reached or even exceeded in many places. Although the analyzed area was small and the measurement was performed under a small load, hardness exceeding 350HV indicates local hardening of material in the HAZ. The results of microindentation tests are shown in Table 4.1.

Strength tests were carried out on round and flat specimens, in accordance with PN-EN 10002-1 standard, which were sampled longitudinally to the axis of plates and the tie member. The analysis was carried out for three specimens taken from each plate.

The test results showed that in terms of strength properties, the material of plates 1 and 3 complies with requirements of PN EN 10025:2002 standard for S355J2G3 steel. The requirements defined in the abovementioned standards in

(a)

(b) (c)

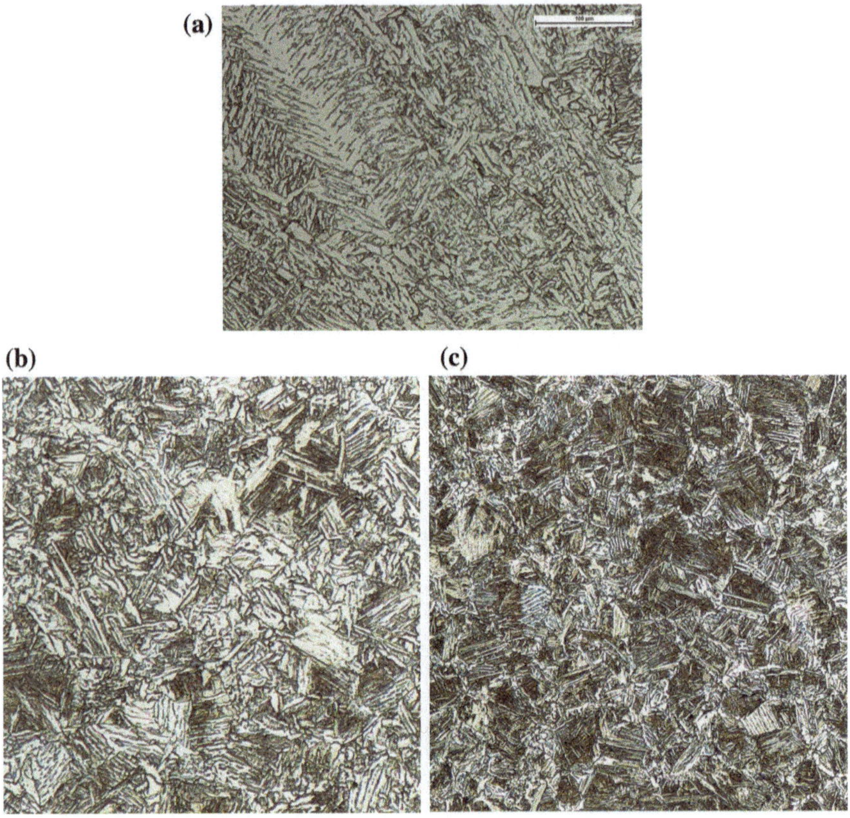

Fig. 4.29 Microstructure of the weld, etched with Mi1Fe, optical microscope: **a** area marked 3 in Fig. 2.68; **b** area marked 2; **c** area marked 4

terms of yield strength are not met by the material in plate 2, in which the steel used had a lower yield strength of Re = 322 MPa than required for the chosen S355J2G3 steel, i.e. Re = 355 MPa. This weakens the cross section of the plate by approximately 10% in terms of ultimate strength and decreases the permissible stress amplitudes during fatigue analysis.

Static and fatigue strength calculations were performed, which demonstrated that the analyzed cross section does not meet the requirements defined in current standards. Material studies indicated structural changes caused by welding defects.

Material studies in the vicinity of welds revealed a ferritic-pearlitic structure with characteristics of the Widmanstätten pattern and with local bands of martensite or bainite, i.e. brittle areas, which were the initiation site of the fatigue fracture. These bands are the consequence of welding defects; for example. from cooling that was too rapid.

It was concluded that the steel used in the bottom plate had lower yield strength of Re = 322 MPa than that required for S355J2G3 steel, i.e. Re = 355 MPa. As a

Fig. 4.30 Microstructure of the area in A-A cross section, Fig. 2.66. The *arrow* indicates the area of probable HAZ of the weld between plate 2 and the bracing. Etched with Mi1Fe, optical microscope

result, the cross section of the plate was weaker by approximately 10% in terms of ultimate strength and the permissible stress amplitudes during fatigue analysis were lower.

The performed shear and fatigue strength calculations demonstrated that the section under test does not meet the requirements defined in current standards. Material studies indicated technological errors in the welded joints and steel properties. The proposed comprehensive testing method allowed for a complete assessment of the causes of the failure and was used to assess its consequences by performing a preliminary dynamic simulation of the complete event.

4.2.2 Failure and Repair of the SRs 2000 Excavator Supports

In contrast to typical supports, the structural solution in the SRs 2000 excavator (Fig. 4.32) enables rotation in three directions. This is possible due to a top fixed bearing and a radial plain bearing on the support column, as well as bearings on the carrying axle of the crawler track in the horizontal direction and laterally in relation to crawler tracks (Fig. 4.33).

The support column transfers vertical and lateral loads between the ring girder and the lower section of the undercarriage, i.e. the crawler tracks and drawbar, and

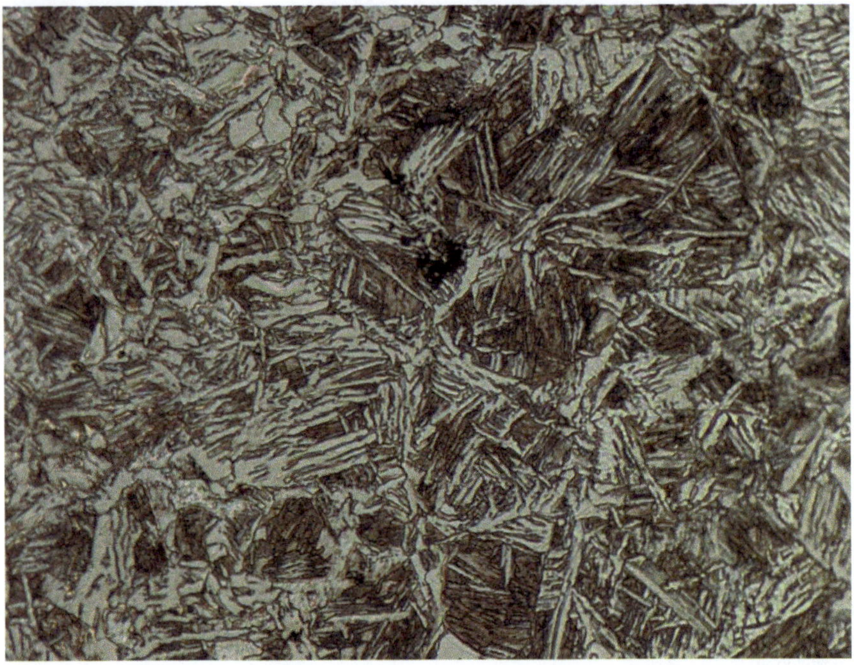

Fig. 4.31 Microstructure of the area in A-A section view, Fig. 4.25. Pearlitic-ferritic structure showing characteristics of Widmannstättena structure and local islands of martensite or bainite. Etched with Mi1Fe, light/optical microscope

Table 4.1 Results of microindentation tests

Indentation no	1	2	3	4	5	6	7	8
Hardness HV0.05	311	349	317	342	379	412	412	387

therefore its condition determines the safety of the machine. It comprises a cylinder (Fig. 4.33) covered with a spherical roller bearing. The moment resulting from the action of lateral forces generated by the undercarriage is transferred by two forces on the spherical roller bearing and below, on the radial plain bearing, and additionally, for component forces in the direction of the drawbar axis, by the restraining moment in the support casing. The lower part of the column consists of a box girder, in which the lower part of the support pipe is mounted.

During overhaul of one of the machines, a fracture was found in the weld under the plain bearing on the support column. The location of the fracture is shown in Figs. 4.34 and 4.35.

Preliminary calculations showed other sites of fracture propagation, i.e. above the pipe band above the fillet weld and on the upper plate of the support box girder (Fig. 4.36), where flaws were also found during additional tests on the machine.

Fig. 4.32 SRs 2000 excavator—general view

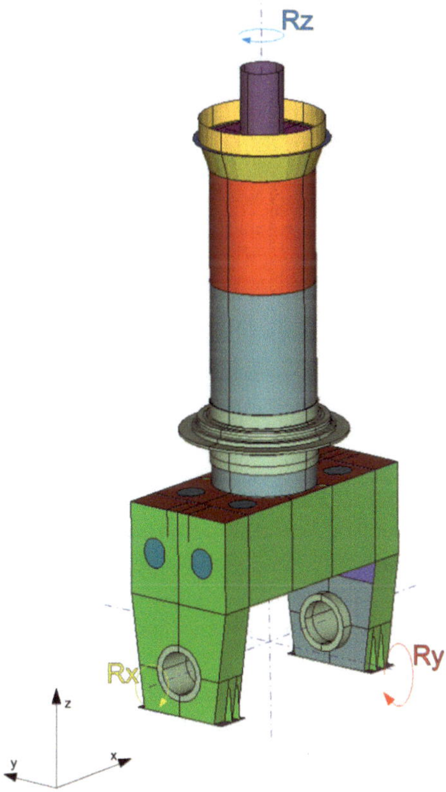

Fig. 4.33 Support column with marked rotation axes

Fig. 4.34 Location of fatigue fractures near the fillet weld under the band

Fig. 4.35 Fatigue cracks under the band on the support pipe

The extensive fatigue damage to the column of the movable support, in the vicinity of two fillet welds connecting the bands under the rings of the horizontal bearing with the column pipe, indicate that there is a structural error, involving the application of fillet welds, and thus the introduction of unacceptable structural notches, welding stresses and structural changes to the material in areas where the bending moments of the column reach near-maximum values. In the vicinity of these welds, the stress amplitudes caused by operational loads exceed the permissible values of fatigue life. This was confirmed by strength calculations performed using the finite element method; detailed results of analyses are presented in [21].

The structural error in this solution was that the slide ring in the bearing is mounted on the vertical column in a faulty manner. In order to ensure that the slide ring of the bearing (with a thickness of 30 mm and a width of 300 mm) does not slide down, the designers decided to weld the ring to the column using a 12 mm

Fig. 4.36 Location of fractures on the top plate of the support box girder

fillet weld, whose only purpose is to hold the ring in place. The fillet weld was made on both sides of the slide ring of the bearing (Fig. 4.37), and since these are field welds, which do not join structural load-carrying elements, they were not classified as welds requiring high strength.

This structural error of making welds to support the slide ring of the bearing on the external surface of the load-carrying column resulted in the cutting of the external longitudinal fibers in the column, and thus introduced a structural notch. This solution substantially decreased the fatigue strength of the column. Strength calculations were performed for the column using FEM [22, 23] for standard loads [19], which indicated that the permissible stress values were exceeded, as shown in Figs. 4.38 and 4.39. The complete strength analysis of the column can is discussed in [21].

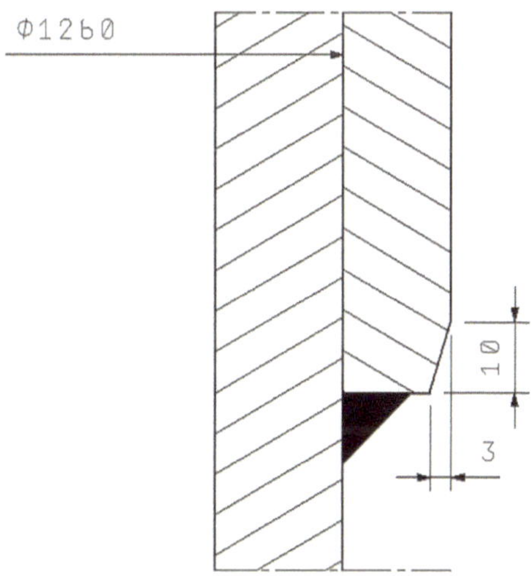

Fig. 4.37 Method of supporting and aligning the slide ring of the plain bearing on the column—incorrect solution due to the fillet weld

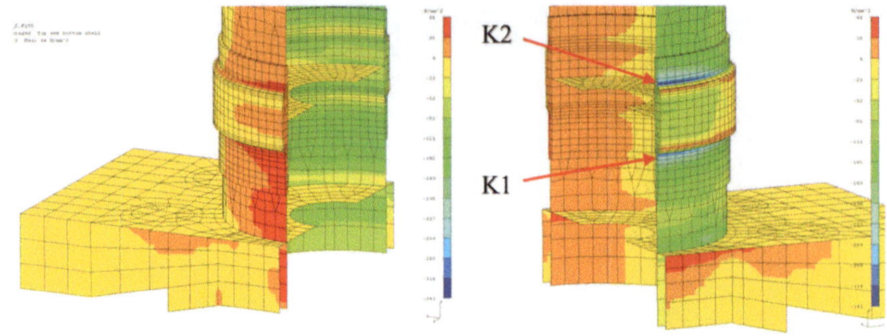

control point	σ_L [MPa]	σ_H [MPa]	σ_a [MPa]	σ_m [MPa]	K
K1	-318	64	**191**	-127	-0,2
K2	-343	54	**198,5**	-144,5	-0,16

Fig. 4.38 Strength calculations for movable supports of the SRs 2000 excavator

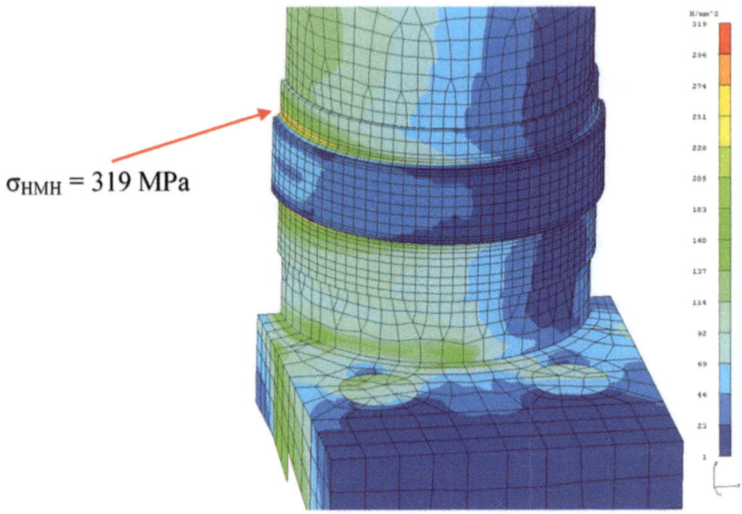

Fig. 4.39 Distribution of HMH contours of equivalent stress in the column

Additionally, the fillet welds were poorly made and had various underfills, microfractures invisible to the naked eye and stress concentrations. Because of all these factors, the internal energy accumulating during many years of excavator operation resulted in a sudden fracture on the grain boundaries of the material and microfractures. This was additionally enhanced by the variable operating conditions of the excavator, which, in the form of load spectra of movable supports, significantly impacted the fatigue strength. The fractures initiated exactly in sites where welds met the parent material of the column. These fractures developed rapidly and

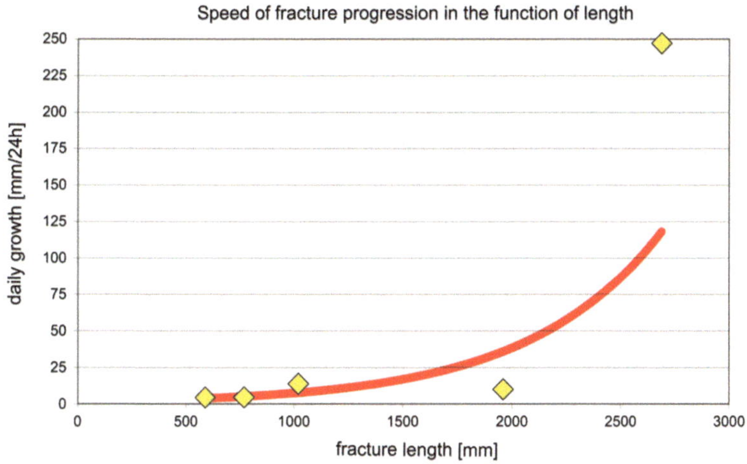

Fig. 4.40 Speed of fracture propagation in columns

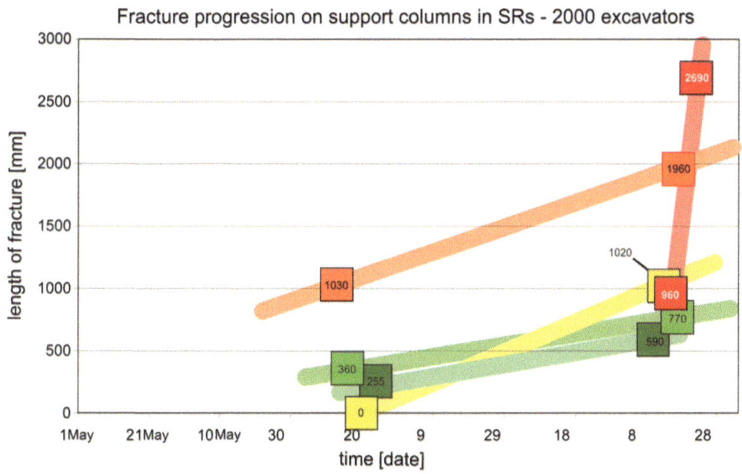

Fig. 4.41 The speed of fatigue propagation in individual columns was rapid in nature

quickly progressed through the column's cross-section, which is illustrated by graphs in Figs. 4.40 and 4.41.

4.2.2.1 Repair Technology

It is recommended that the final repair of the movable support column near the horizontal plain bearing is made by changing the structural shape of the column's segment. A segment of the column with a 36-mm-thick wall above the support's box girder will be replaced by a pipe with a variable-thickness wall along the column length, as shown in Fig. 4.42.

Technologically, the final repair of the column involved the replacement of the column section via forging, as shown in Fig. 4.42. This repair technology ensures that the structural solution will meet the fatigue strength criteria. The forging will be connected with the column by means of two groove welds, tested over the entire circumference.

The solution with the forging for the movable support column was tested using FEM strength calculations, whose results are fully analyzed in [21]. Figure 4.43 shows an example of the results.

The forging additionally has a mounting for the plain bearing. This solution ensures 100% reliability and safety. The columns in all damaged machines were repaired using the presented technology. Figures 4.44 and 4.45 are examples of such repairs.

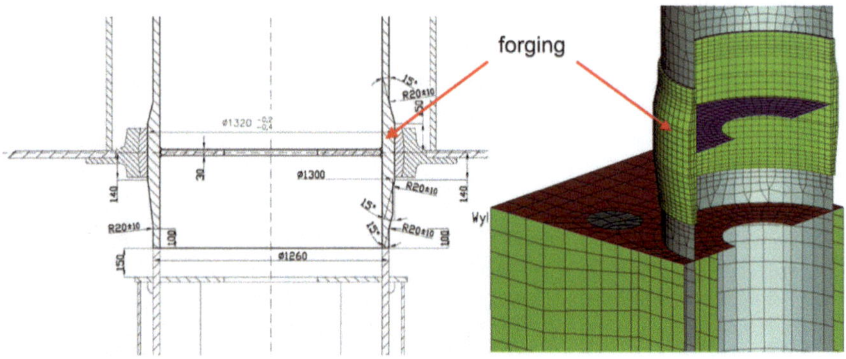

Fig. 4.42 Outlines of the pipe section for final repair connected with a butt weld

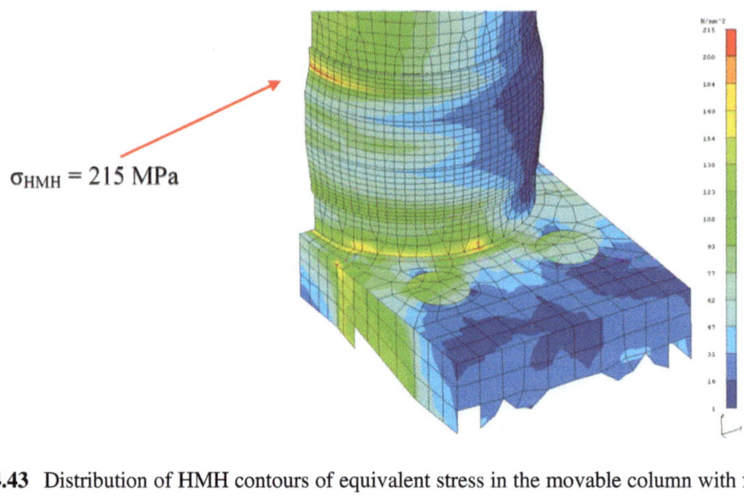

Fig. 4.43 Distribution of HMH contours of equivalent stress in the movable column with forging

Fig. 4.44 Support column of the SRs 2000 excavator repaired using the Wrocław University of Technology method, *front view*

Fig. 4.45 Support column of the SRs 2000 excavator repaired using the Wrocław University of Technology method, *side view*

4.2.3 Fracture of a Half Axle of Crawler Track Set in the ZGOT-11500.100 Stacker

During operation in winter conditions, the steerable track of the stacker failed (Figs. 4.46 and 4.47). The stacker driving chassis consists of three sets of crawler tracks, one of which is steerable and two that are not steerable. A crawler set comprises two crawler track frames, a steering frame and bogies. Figure 4.48 shows a diagram of the steerable crawler track set.

The crawler track frame and the steering frame are box structures built of 25–30 mm steel plates. The span of track frames is 5.9 m, whereas the distance between the steering shaft and the axle of track frames is 8.5 m. The analyzed

Fig. 4.46 ZGOT-11500.100 stacker

Fig. 4.47 View of the steerable crawler track after failure

Fig. 4.48 Diagram of the steerable crawler track

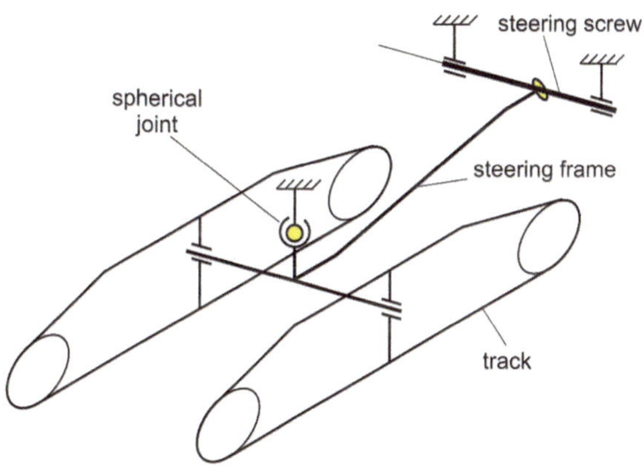

stacker has two half-axles connecting the track frames with the bogie, i.e. it does not have a single axle. Such a solution is often used to decrease mass. It should be noted that half-axles in such machines are about 3 m long and have a diameter of approximately 520 mm, tapering down to 300 mm inside the track frame, which gives a mass of approximately 3800 kg. This solution, however, requires a strong support and a mechanism that prevents the axles from sliding out. Figure 4.49 shows the method of mounting half-axles inside the track box frame.

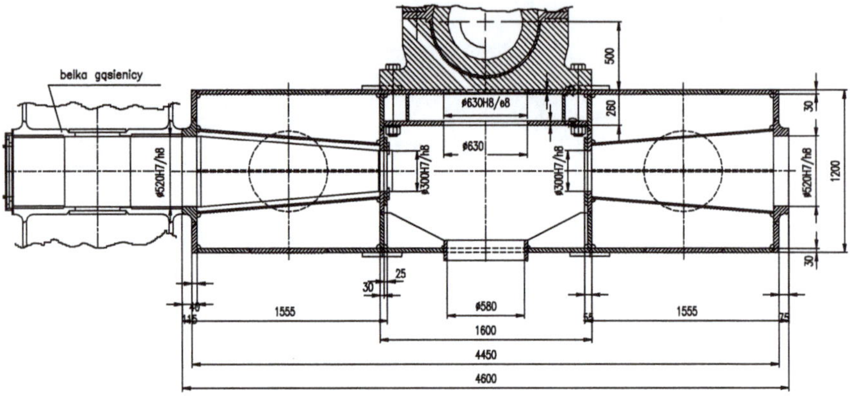

Fig. 4.49 Method of mounting half-axles inside the track frame

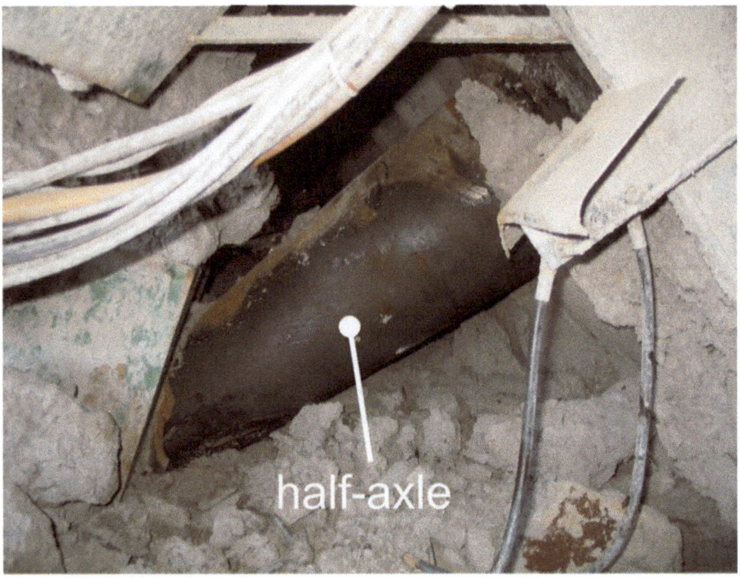

Fig. 4.50 Half-axle of the steerable track after sliding out

The failure occurred (Fig. 4.51) when one of the half-axles slid out after its end broke, which is shown in Fig. 4.50.

After lifting the stacker portal and the superstructure of the machine and after pulling out the track frame, it was possible to inspect the damaged end of the half-axle. Figure 4.52 shows the damaged end of the half-axle and the chipped pieces.

Fig. 4.51 Effects of the failure of the steerable crawler track in the stacker

In order to identify the causes of this event, the following tests were performed:

- Defectoscopy of the track frame and half-axle,
- Geometry measurements of the track's steel plates and geometry measurements of the half-axle,
- Metallographic examination of the half-axle material,
- Analysis of forces acting on the steerable crawler track and the whole undercarriage [24] in accordance with the current standards [19],
- Strength calculations using the finite element method for the steerable crawler track and the steering frame.

Another issue was the lifting of the stacker in order to repair the damaged crawler track. To ensure safety, in this case a series of FEM numerical analyses were performed in order to verify the method of lifting the machine.

The general principles of calculations for tracked undercarriages are discussed in [25], which analyzes the individual components of travel resistance forces in detail, i.e.:

- Internal force of track movement,
- Deformation force of foundation,
- Lifting force,
- Travel resistance on a curve.

An analysis of these forces helps to define the loads acting on individual elements of the crawler track set. The analysis of the causes of failure focused on the analysis of forces acting on the crawler tracks traveling round a curve.

(a)

(b)

Fig. 4.52 **a** Damaged end of the half-axle in the steerable track; **b** the chipped pieces

On the other hand, in accordance with the current standard DIN 22261-2 on calculations for surface mining machines and the corresponding standard PN-47000-2 for tracked undercarriages, calculations should be performed for the following cases:

- Traveling round a curve (L),
- Lateral slip of the track (QQ),
- Crawler blockage (LL), e.g., in the case of drive failure in one of the tracks.

These calculations also include the loads exerted by the superstructure on a slope and under wind pressure. Safety systems installed in surface mining machines do not allow a turning maneuver to be performed while stopped. Turning is only possible while the machine is in motion. Figure 4.53 shows an example diagram of loads acting on the crawler track set when travelling round a curve.

The following section describes the calculation procedure using the finite element method [22, 23], which was essential in verifying whether the sections for load-carrying elements were chosen correctly. Calculations were carried out on a shell-beam model for the box structure of the track frame with the steering frame, and on a solid model for half-axle analysis.

After setting the forces for individual load cases, numerical calculations were performed. Figures 4.54 and 4.55 show examples of Huber-Mises equivalent stress distribution for the case involving traveling round the curve.

The analysis of the state of stress for all required load cases indicates that standard permissible stress was not exceeded [19]. Relatively high stress was observed only in the vertical ribs of the half-axle casing (Fig. 4.55).

In the next stage, the loads acting on the half-axle were analyzed, in particular on the end of the half-axle, where a lock was installed preventing the half-axle from sliding out. When traveling round a curve, substantial axial forces are generated by the lateral shear of the ground. In unfavorable circumstance, these forces can add up to the lateral forces generated by tilting and the wind pressure forces. Figure 4.56 shows the method of protecting the half-axle from sliding out, whereas Fig. 4.57 shows the broken end of the half-axle, which was the cause of stacker failure.

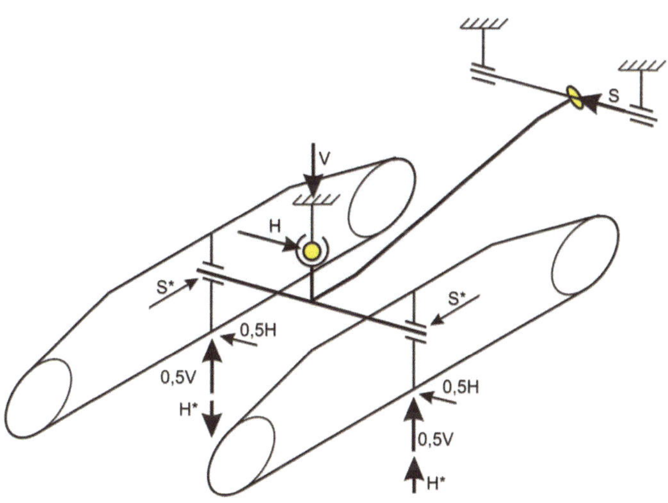

Fig. 4.53 Forces acting on the crawler track set while traveling round a curve

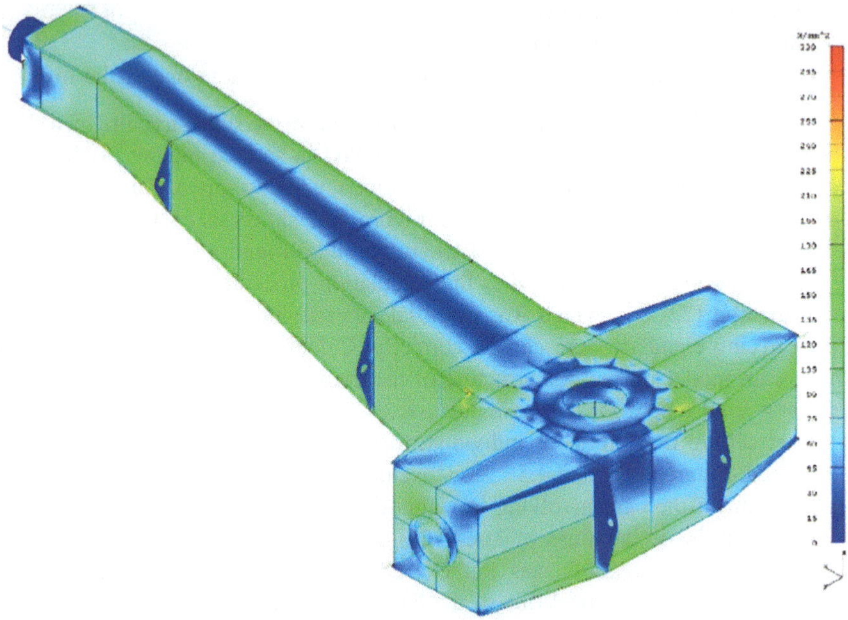

Fig. 4.54 Distribution of equivalent stress in the steering frame of the stacker

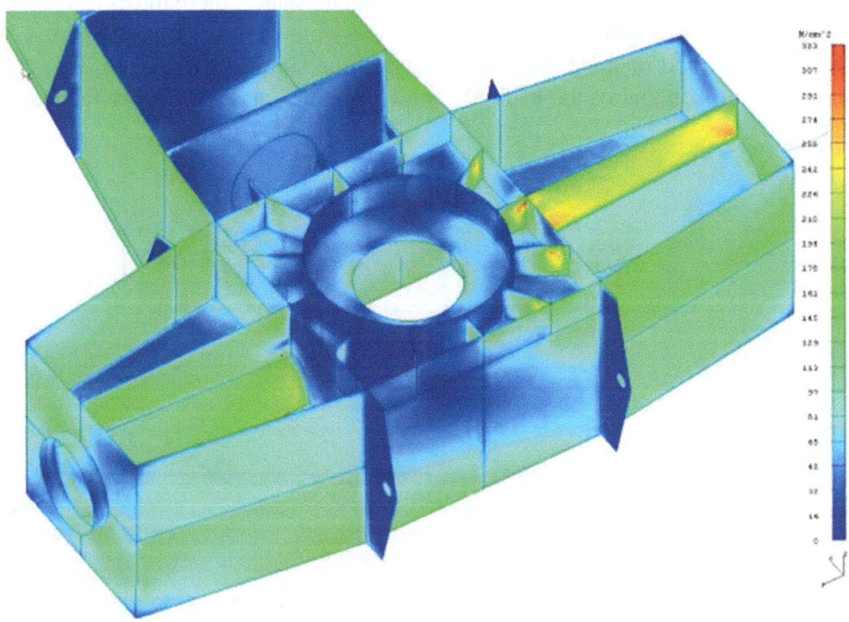

Fig. 4.55 Distribution of equivalent stress in the steering frame of the stacker after uncovering the top plate

Fig. 4.56 Method of preventing the half-axle from sliding out

Fig. 4.57 The broken end of the half-axle

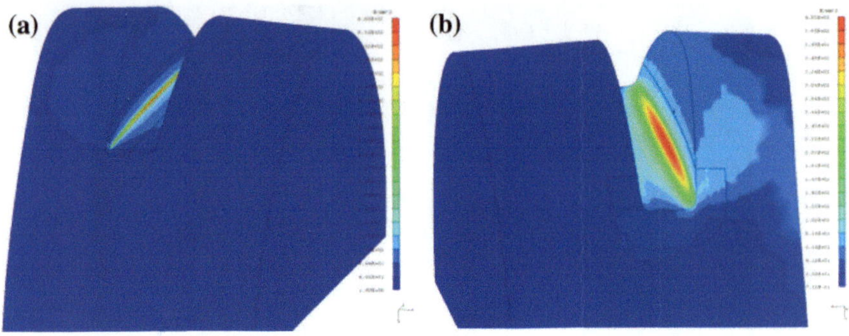

Fig. 4.58 Distribution of stresses in the groove for locking collar against sliding out (non-linear elastic and plastic calculations): **a** $r = 0.2$ mm; **b** r = 4 mm

Considering the most unfavorable loading scenario, FEM calculations were performed for the end of the half-axle. It should be noted that the bottom of the groove for protective plates had a very small curvature radius, which causes substantial concentration of stress under axial loads. A measurement of the cross-section revealed that the radius was only $r = 0.2$ mm.

Figure 4.58 shows the distribution of stress obtained through FEM analysis for the radius values of $r = 0.2$ mm and $r = 4$ mm, the latter being used for comparison purposes, whereas Table 4.1 shows the maximum values on the bottom of the groove for two values of the axial force transferred to the locking collar. These values result from the adopted mechanisms of lateral force transfer, i.e. either entirely by the lock or partially by the bearing journals, with consideration of the friction coefficient in both cases.

The results of FEM calculations indicate that when axial forces act on the bottom of the groove with a radius of $r = 0.2$ mm, the stress substantially exceeds the plastic limit. When the assumed radius of $r = 4$ mm was introduced, the stress was lower than the plastic limit.

In the next stage, metallographic examinations were performed on three elements, sampled from the broken axle locking collar of the track half-axle. The obtained elements are marked 1, 2, and 3 (Figs. 4.59 and 4.60). Macroscopic examinations were performed with the naked eye and using a stereoscopic microscope with magnifications up to 30x. The chemical composition of the half-axle material was analyzed using the gravimetric method for the following elements: C, Mn, Si, P, S, Cr, Mo, Ni. Microscopic examinations were also performed.

Macroscopic examinations revealed the occurrence of fractures that originated in the locking collar of the track half-axle. Fractographic examinations showed that the resulting fractures resemble a fatigue fracture with a coarse-grained brittle shear zone (Figs. 4.61 and 4.62). It was determined that both fractures were initiated in the area of cross-section change. The fatigue zone is located along the undercut sites and is indicated by symbols K_1 and K_2 in Fig. 4.60.

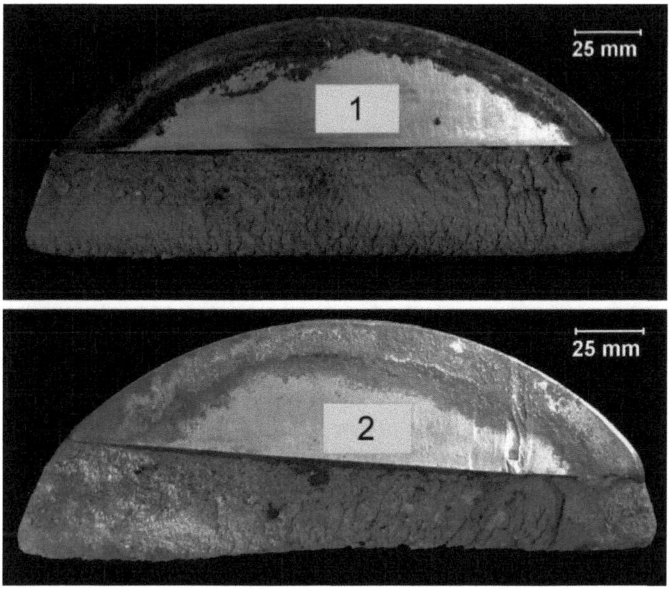

Fig. 4.59 General view of pieces of the broken track half-axle—elements 1 and 2

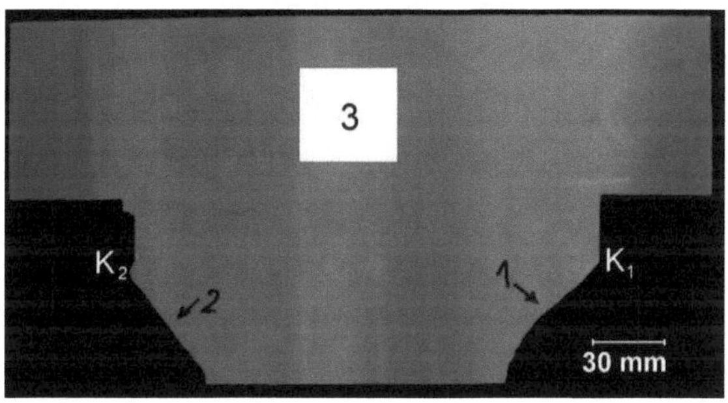

Fig. 4.60 General view of the cross-section of a piece of the broken half-axle—element 3. *Arrows* mark the contact surfaces of fractures in damaged pieces 1 and 2. Symbols K_1 and K_2 mark the location of undercuts, along which the fatigue zones of fractures are located

The wear marks on the contact surface indicate that surface pressures were distributed around the locking collar circumference, and not across its entire surface. Such a distribution of pressure stress is the result of the oval shape of the fatigue zone, which is due to the fact that, with pressure forces distributed evenly along the circumference, the resulting maximum bending moments occurred in the central part of the locking collar, in the vicinity of points S_1 and S_2.

The area marked "a" in Fig. 4.63 is the site identified using macroscopic examinations, from which a sample was taken for further metallographic examinations.

From both sides of the "a", sample measurements were taken of the groove edge radius in the locking collar, marked N in Fig. 4.63, into which safety plates were

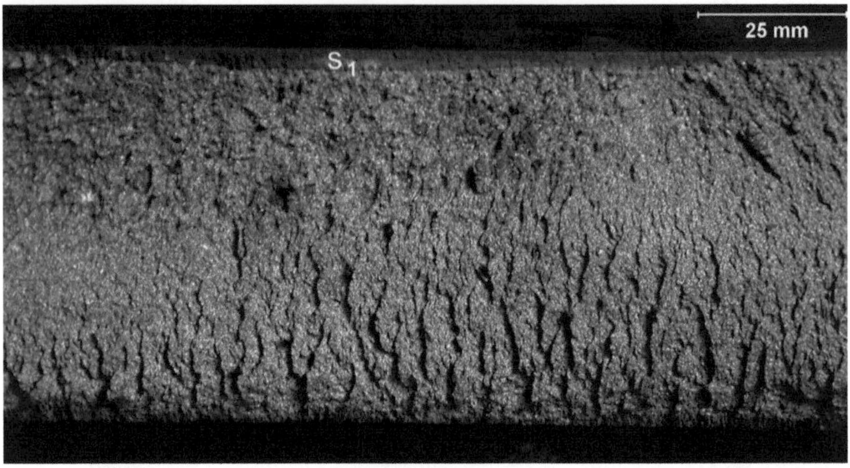

Fig. 4.61 View of surface of fracture 1 with marked edges. The S_1 symbol marks the fatigue zone of the widest fracture

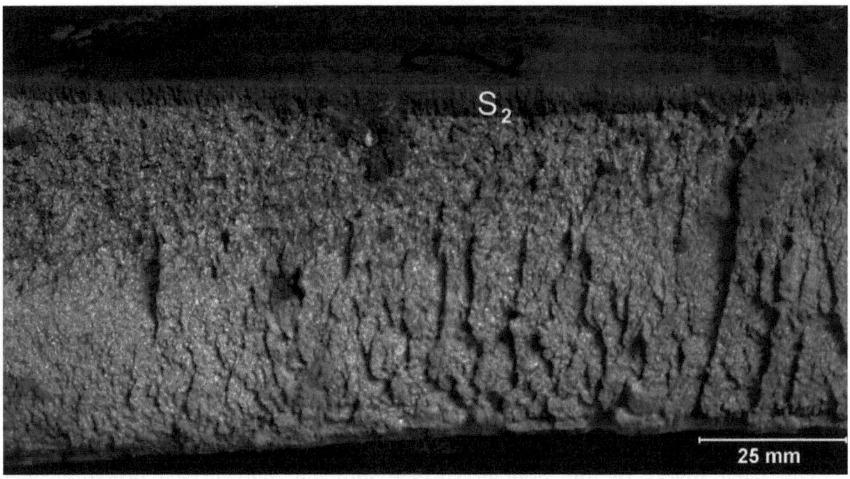

Fig. 4.62 View of surface of fracture 2 with marked edges. The S_2 symbol marks the fatigue zone of the widest fracture

inserted. Curvature radius R equaled 0.2 mm (Fig. 4.64). This radius value was used in FEM calculations, whose results are presented above.

Examinations also revealed a fracture that originated on the edge of the groove and that progressed parallel to the edge of cross-section, at a 45° angle to the shaft axis and along the ferrite and pearlite grain boundaries (Fig. 4.65).

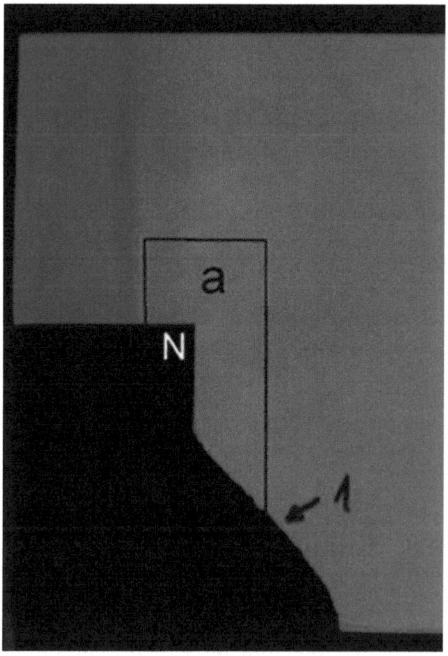

Fig. 4.63 Site from which a sample was taken for further metallographic examinations, identified using macroscopic examination

Fig. 4.64 R 0.2 curvature radius of the locking collar groove edge on the left side of the sample

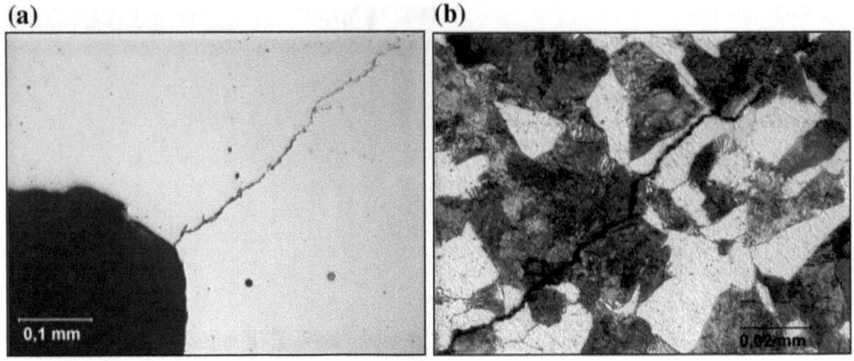

Fig. 4.65 Groove edge in the locking collar using a light microscope: **a** visible fracture originating in the area of cross-section change, without etching; **b** magnified detail of the area after etching, visible fracture running along the ferrite and pearlite grain boundaries

4.3 Summary

Based on FEM analyses of the crawler track set, and of the half-axle end, it was concluded that the main cause of failure was a design flaw in the method of protecting the half-axle from sliding out.

Material studies revealed that the fractures in the track support half-axle lock resemble a fatigue fracture, i.e. the shear zone was brittle and coarse-grained. Fatigue zones originated in the area of cross-section change. They progressed from the edge of the groove through the material at an angle of approximately 45° to the symmetrical axis of the half-axle and were oval-shaped, with the largest width near the center of the groove edge.

The low ratio of fatigue zone to shear zone indicates the presence of substantial real stress in the structure or the application of very low safety coefficients.

Microscopic examinations also revealed fractures (Fig. 4.65), which indicate the initiation of a fatigue fracture. Fractures originated from the area of cross-section change in the lock and progress through the material at an angle of approximately 45°, parallel to the entire edge of the sample taken from the supplied element. They were very similar to the damage that caused the half-axle protection to break.

The aforementioned examinations show that the method of protecting the axle from sliding out was not erroneously designed and constructed. The failure was directly caused by the excessively small curvature radius on the bottom of the groove, which was the point of origin of the fracture, and caused the half-axle to extend and, consequently, the stacker to fail.

It was proposed to change the protection by increasing the curvature radius on the bottom of the groove to $r = 5$ mm. In the present solution, the protection is

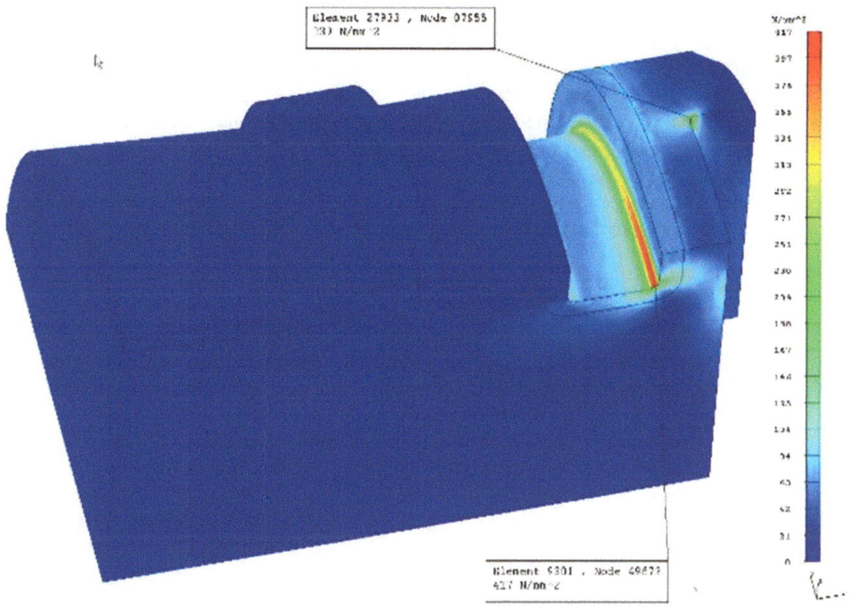

Fig. 4.66 Distribution of stress in the new solution of the half-axle end

effective around the entire circumference. Additionally, the diameter of the half-axle in this place was increased from 300 to 400 mm. FEM calculation results show that even under maximum force of $F = 1393$ kN, the stresses at the bottom of the circular groove in the new solution are equal to $\sigma = 417\ \text{N/mm}^2$, which increases the safety coefficient to more than two (Fig. 4.66).

References

1. Kasztelewicz Z (2012) Bucket wheel excavators and spreaders. Technology of operation, Art-Tekst, Kraków
2. Babiarz S, Dudek D (2007) Annals of failures and catastrophes of surface mining machines in Poland. Wroclaw University of Technology Publishing House
3. Bošnjak S, Zrnić N, Simonović A, Momčilović D (2009) Failure analysis of the end eye connection of the bucket wheel excavator portal tie-rod support. Eng Fail Anal 16(3):740–750
4. Savkovic Mile, Gašic Milomir, Arsic Miodrag, Petrovic Radovan (2011) Analysis of the axle fracture of the bucket wheel excavator. Eng Fail Anal 18:433–441
5. Rusiński E, Czmochowski J, Pietrusiak D (2012) Problems of steel construction modal models identification. Eksploatacja i niezadowność—Maintenance and Reliability. 14(1):54–61
6. Augustynowicz J, Dudek D (2002) Problems of the instability states in machine. Syst: J Trans Syst Sci 7(1):10–16
7. Bialas K (2010) Passive and active elements in reduction of vibrations of torsional systems. Solid State Phenom

8. Glowacz A (2014) Diagnostics of DC and induction motors based on the analysis of acoustic signals. Meas Sci Rev 14(5):257–262
9. Gottvald J (2012) Analysis of vibrations of bucket wheel excavator Schrs1320 during mining process. FME Trans 40(2012):165–170
10. Jabłoński M, Ozga A (2012) Determining the distribution of values of stochastic impulses acting on a discrete system in relation to their intensity. Acta Physica Polonica. A 121(1): A-174–A-178. ISSN 0587-4246
11. Smolnicki T, Kowalczyk M, Pietrusiak D (2010) Identification of dynamic characteristics of the stacker. In: 6th international conference mechatronics systems and materials, Opole, 2010, pp 1–8
12. Rusiński E, Cegiel L, Michalczyk A, Moczko P, Olejarz J, Pietrusiak D (2015) Investigation and modernization of buckets of surface mining machines. Eng Struct 90:29–37. doi:10.1016/j.engstruct.2015.02.009
13. Bocian M, Jamroziak K, Kulisiewicz M (2014) An identification of nonlinear dissipative properties of constructional materials at dynamical impact loads conditions. Meccanica 49 (8):1955–1965
14. Rusiński E, Czmochowski J, Iluk A, Kowalczyk M (2010) An analysis of the causes of a BWE counterweight boom support fracture. Eng Fail Anal 1(17):179–191
15. Gottvald J, Kala Z (2011) Variance-based sensitivity analysis of tangential digging forces of the bucket wheel excavator SchRs 1320. In: Proceedings of the international conference on recent researches in engineering and automatic control, Puerto De La Cruz, Spain, 2011, pp 204–209
16. Hadjigeorgiou J, Poulin R (1998) Assessment of ease of excavation of surface mines. J Terramechanics 35(3):137–153
17. Pietrusiak D, Moczko P, Czmochowski J (2013) Field and numerical testing of the BWE SchRs4600. 50 dynamic behavior. In: Proceeding of the IMAC XXXI, Garden Grove, USA, 11–14 Feb 2013
18. Zaveri K, Phil M (1985) Modal analysis of large structures—multiple exciter systems. Bruel & Kjaer
19. DIN 22261-2 Bagger, Absetzer und Zusatzgeräte in Braunkohlentagebauen
20. Czmochowski J (2008) Identification of modal models of excavators in lignite mining. Wroclaw University of Technology Publishing House
21. Technical Report IMDO. PWr. 2007, Ser. SPR ; no 4
22. Rusiński E, Czmochowski J, Smolnicki T (2000) Advanced finite element method in load carrying structures. Wroclaw University of Technology Publishing House
23. Hughes T (2000) Finite element method. Dover Publications
24. Maślak P, Smolnicki T, Pietrusiak D (2013) Strain gauges measurements and FEM analysis of elements of chassis of open cast mining machines. Tehnicki vjesnik—Technical Gazette 20(4):655–658
25. Durst W (1986) Schaufelradbagger, Trans Tech Publications

Chapter 5
Modernization of Existing Machines

5.1 Objectives of Modernizations—Increase in Lifetime of Machines and the Optimization of Operational Parameters

Since the majority of surface mining machines that operate in Polish mines are several years old—most of them even several decades old—it is necessary to modify and modernize them. These modernizations mainly stem from the imperfection of their structure. Surface mining machines are prototypical structures. The unsuitability of their structure to the operating conditions and many other weaknesses are discovered only after using them in real life situations. Therefore, sometimes modernizations are required immediately after the machine is put into operation. In most cases, however, the purpose of modernizations is to extend the lifetime of machines that have been used for many years. Due to the changing geological conditions and the occurrence of various hard-mineable formations in the excavated material, and owing to the very nature of the digging process and operation under difficult conditions, elements of machines are subject to rapid degradation. The method of assessing technical conditions helps to identify those elements that are particularly worn. Such elements undergo a structural analysis and a decision is made to either repair/replace them or to modernize them in order to extend their durability. Another purpose of modernizations is to improve the operating parameters of the machine. This also includes those parameters that directly translate into economic benefits, i.e. productivity and uptime. An increase in uptime also occurs if modernization of the load-carrying structure improves the reliability of the machine. In many instances, modernizations are also introduced in order to improve the ergonomics, comfort and safety of operation. These parameters indirectly affect the productivity and uptime of the machine and thus the purely economic aspects. Increasingly, with the advancement of technology, modernizations no longer consist of improving or altering the existing design or machinery elements, but in introducing an entirely new solution. Today, the standard

E. Rusiński et al., *Surface Mining Machines*,
DOI 10.1007/978-3-319-47792-3_5

equipment of mega machines already includes pneumatic vibroisolated suspension of operator cabins, mechatronic overload clutches or monitoring systems, but such features were not implemented in structures that were put into use only a few years ago.

Many of the introduced solutions have been created using modern designing techniques based on numerical methods and experimental tests with the application of specialized equipment. Most of these techniques were not in common use at the time when surface mining machines that currently operate in Poland were designed. Today, they are successfully used to improve many operational and design parameters.

Section 5.2 is an overview of chosen modernizations of existing machines as well as entirely new solutions. Section 5.3 presents a case study that will be used to illustrate the comprehensive approach and the level of effort required to introduce a seemingly slight change in the number of buckets on a bucket wheel.

5.2 Modernizations—Examples of Solutions

Load-carrying structures of surface mining machines are usually modified in order to improve their durability or eliminate problems related to dynamics, and, less frequently, to improve ergonomics or to increase productivity. However, in the latter case, such modifications are often an intermediate step required in order to introduce new units (e.g., the excavating unit), drives or conveyor belts, whose modernization directly increases the productivity.

However, at a time when people are searching for solutions that minimize the operating costs of surface mining machines [1, 2] and eliminate the likelihood of failures [3–5], it is becoming increasingly important to optimize the technical and operational parameters related to the mining of overburden and coal in surface mines. Such attempts are being made, among others, in the field of excavating units of bucket-wheel excavators and bucket-chain excavators. The primary objectives of these projects include the following:

- To increase the durability of excavating units,
- To decrease maintenance activities,
- To increase the efficiency of excavation,
- To reduce the dynamic effects related to excavation,
- To increase the uptime of machines.

5.2.1 Modernization of Buckets

One of the elements that has recently become subject to modernizations is buckets. Bucket-wheel excavators usually have buckets with a closed shell and a cutting

Fig. 5.1 Examples of excavator buckets

edge with corner cutters. There are also bucket solutions with teeth protruding from the cutting edge (Fig. 5.1).

It is important for the buckets to be easily removable during use. The procedure of bucket replacement is illustrated in Fig. 5.2.

Once the cutting elements of the bucket get worn, they should be replaced, which is possible in a workshop after detaching the bucket from the bucket wheel. It is crucial to use attachment systems that allow the bucket to be replaced quickly. Figure 5.3 illustrates an innovative solution (patent application no. P404090) that enables the bucket to be replaced much easier and quicker than before.

The durability of the bucket is also very important. Even in systems that enable quick and easy bucket replacement, the buckets will not be of much use if they have to be replaced too often. Therefore, research is conducted into optimum adjustment of buckets to match the parameters of a given machine. It is equally important for the teeth on the buckets to enable long-term and fault-free operation without having to be replaced frequently.

The requirements imposed by the mining industry on teeth manufacturers can be summarized into the following conditions:

- The tooth must be abrasion-resistant (it must be in service for a given number of hours),
- During use (wear), the tooth must simultaneously self-sharpen,
- During use, the tooth cannot break, i.e. its core must be flexible, which requires appropriate heat treatment.

Fig. 5.2 Bucket replacement

Fig. 5.3 Bucket attachment system—patent application no. P404090

In order to modernize the existing buckets or develop entirely new solutions, it is necessary to conduct a wide range of activities, whose aim is to assess the new or existing structures and to introduce solutions that could improve the technological and operational properties of these buckets. These activities include the following:

- Testing the actual loads on excavating units,
- Analyzing the trajectory of cutting elements of the bucket,
- Analyzing the distribution of loads and optimizing the geometry of the cutting elements with the application of numerical methods,
- Performing strength analysis using numerical methods,

- Operational tests involving the measurement of bucket wear, operating time, energy parameters (consumption of energy), performance, excavating parameters (excavating speed), etc.,
- Economic analyses resulting from the application of existing and new solutions.

Owing to the analysis of loads by measuring energy consumption as a function of physical wear of buckets (decrease in the length of cutting elements), it is possible to identify the bucket, which is characterized by the lowest energy consumption for a given machine. This also offers the possibility to test new structural solutions and to compare them with solutions that are already in use in order to determine whether they are appropriate. The graph in Fig. 5.4 shows the relationship between energy consumption and wear of teeth on the bucket.

Other research involves the analysis of movement trajectories of the bucket's cutting elements in order to assess whether the bucket is compatible with the machine on which it is to be mounted. Basic analysis of cutting kinematics [6–8] involves only the geometry of the excavated lump without going into detail regarding the positioning of cutting elements on the cutting edge. This can cause the overlapping of "teeth marks" of subsequent buckets and thus lead to the increase in loads acting on adjacent teeth and to the output loss.

It is known that the resultant component of the bucket's velocity vector is the horizontal vector of linear velocity resulting from the rotation of the superstructure combined with the perpendicular vector of linear velocity resulting from the rotation of the bucket wheel. Both these vectors move tangentially in relation to the circles, whose radiuses are equal, respectively, to the distance from the center of the axis of superstructure rotation to the furthest edge of the bucket, and from the center of the rotation axis of the bucket wheel to the furthest edge of the bucket. The

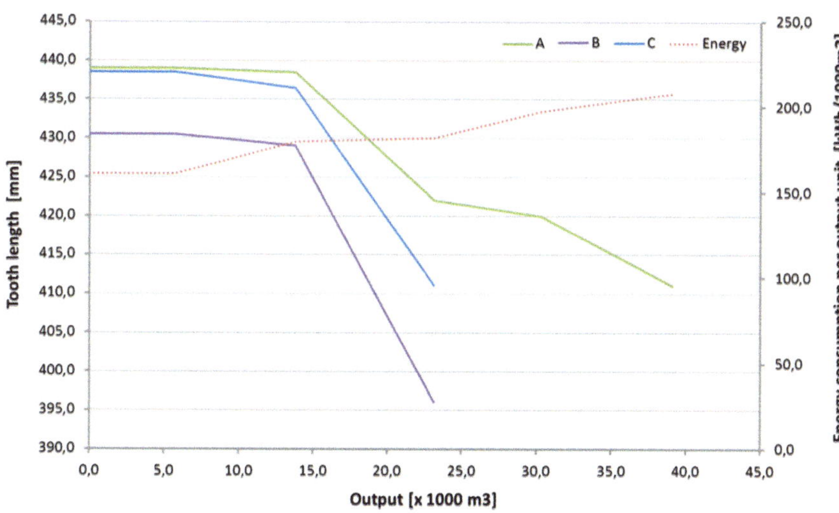

Fig. 5.4 Relationship between energy consumption and the wear of teeth on the bucket

resulting movement trajectory is a 3D curve (Fig. 5.5). Naturally, such a situation occurs when we disregard the possible movement resulting from the lowering of the excavating boom and from the travel of the machine.

In the case of a bucket-chain excavator, instead of a rotating movement of the bucket wheel, there is the movement of the chain along the boom. In combination with the rotation of the superstructure, the resulting curve is 2D (Fig. 5.4), as opposed to the three dimensions in the case of the bucket-wheel excavator (this pertains to the trajectory only in a certain range of the movement).

Moreover, the obtained curves can be used to analyze the movement trajectory of a single bucket tooth. The objective of such an analysis is to determine whether at a given velocity of superstructure there will be a situation in which the movement trajectories of individual teeth on the bucket overlap, as shown in Fig. 5.6 (which shows overlapping of trajectories of the first and second tooth located on the front of the cutting edge). As shown in [9], the analysis of only the first pair of teeth is the most important, because after several passes of buckets, previous teeth marks are obscured. It is a fact, however, that the higher the velocity of superstructure rotation, the further apart the trajectories of individual buckets are, which implies that other pairs of teeth should also be analyzed.

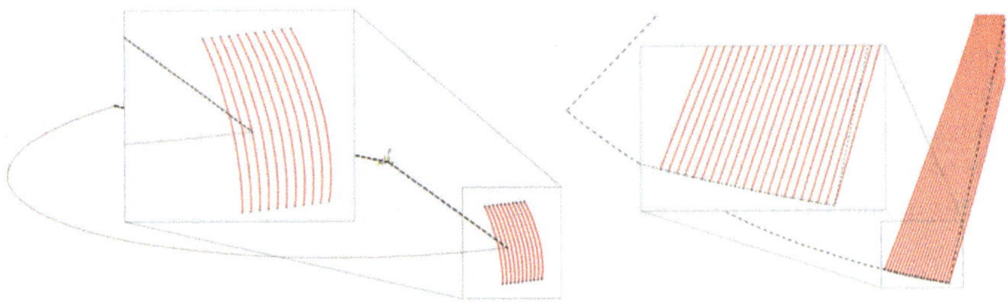

Fig. 5.5 Examples of trajectories of buckets in bucket-wheel and bucket chain excavators

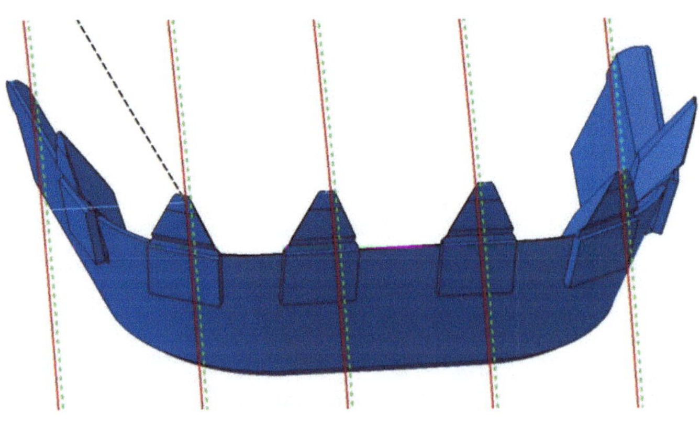

Fig. 5.6 Overlapping of trajectories of the first and second tooth

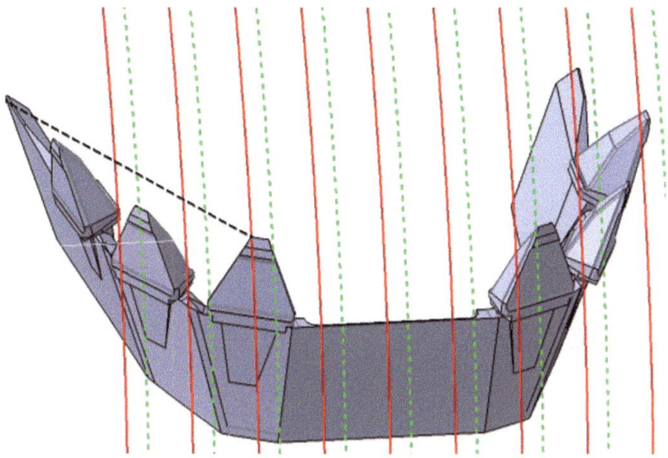

Fig. 5.7 Cutting trajectories of tooth 1 and 2 after a change in the rotation velocity

If the geometry of the cutting tool cannot be modified, there is the possibility to select the range of rotation speed of the superstructure and thus eliminate the overlapping of cutting trajectories of teeth (Fig. 5.7).

By performing the aforementioned analyses at the stage of modernization or design of a new bucket, it is possible to introduce geometric changes that will improve the excavation performance.

Of equal importance is the analysis of loads resulting from the cutting process itself. By applying modern numerical analyses, it is possible to run simulations and study phenomena that could not have been studied before.

Numerical simulations allow for a broad-scope analysis of mechanical phenomena. Among others, it is possible to analyze the loads acting on the cutting elements of the buckets (teeth, corners) during excavation. To achieve this, the finite element method or the discreet element method is used [10, 11]. An example of such an analysis for bucket teeth used in KWK1200 and KWK1500 excavators is presented below.

In order to evaluate the chosen cutting angles for buckets, a numerical analysis was carried out for the distribution of forces acting on the cutting tool. Loads were evaluated using the coefficient (1)

$$S_{\mathrm{k}} = Ft/Fr, \tag{5.1}$$

which is the relationship between the following component forces acting on the cutting tool (Fig. 5.8):

Ft tangential component to the excavating circumference,
Fr radial component to the excavating circumference

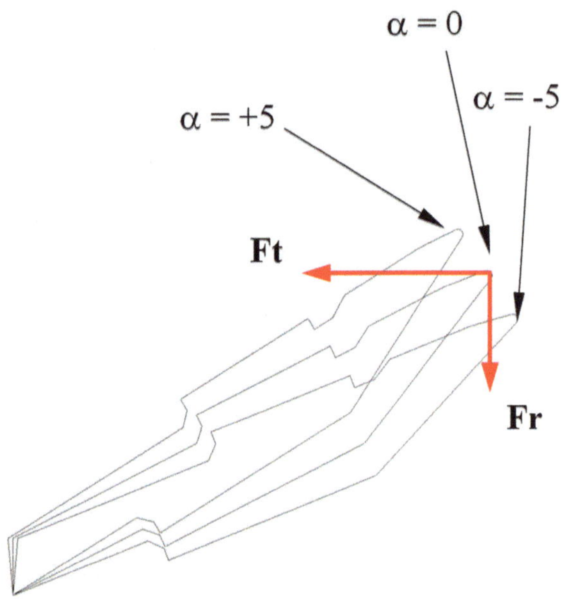

Fig. 5.8 Components of forces and position of the cutting tool in an analysis of distribution of loads

The S_k coefficient was calculated for three values of angles in relation to the current real angles:

- −5—inclination of tooth by −5 degrees to the "inside" of the bucket, in relation to the current angle (sectional),
- 0—current real inclination,
- +5—inclination of tooth by +5 degrees to the "outside" of the bucket, in relation to the current angle (sectional).

The result of analyses is the variability of the S_k coefficient as a function of the contact angle for the analyzed cutting tools. The results are displayed as a graph in Fig. 5.9.

The Ft/Fr coefficient helps to establish:

- Susceptibility of the tooth (corner cutter) to "breaking" to the inside of the bucket—the threshold value is 1,
- Susceptibility of the tooth (corner cutter) to "breaking" to the outside of the bucket—increases with the increase in the coefficient value,
- Ability of the tooth (corner cutter) to self-sharpen—increases with the increase in the coefficient value,
- Required energy—increases with the increase in the coefficient value.

Eventually, a strength analysis is also performed using the finite element method (FEM). This allows for the optimization of the geometry of the entire bucket and not only its cutting edges.

The strength and durability of elements in the excavating unit of surface mining machines is the main determinant of their faultless operation. Bucket designs are

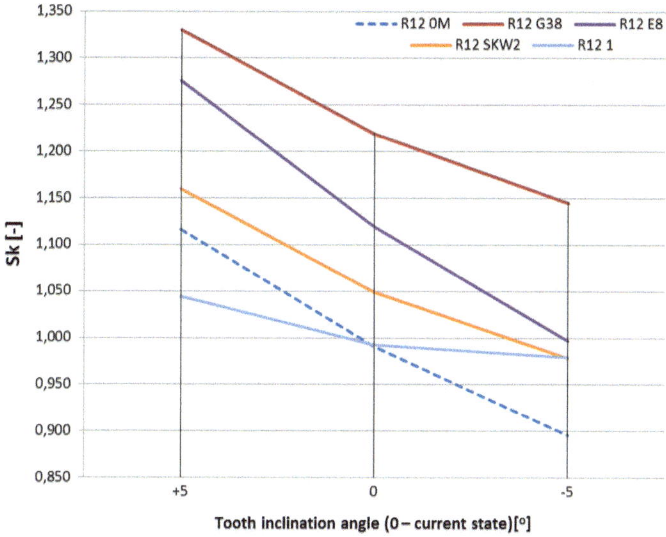

Fig. 5.9 Variability of the S_k coefficient for the buckets used in KWK1500 and KWK1200 excavators

usually based on criteria related to ultimate and buckling strength, particularly if they are used to excavate overburden with non-mineable formations. In the case of materials with constant excavating resistance (mainly coal), due to long-term operation, the bucket design also takes into account fatigue strength. Such analyses should be performed using numerical methods (e.g., FEM), which help to identify the state of stress and deformation at every point of the analyzed structure. It is also possible to design such a bucket structure that would enable damage control in the case of a critical overload of the excavating unit. Therefore, the bucket can also help to protect the surface mining machine against overload. Examples of simulation results are shown below.

Loads that determine the bucket strength are directly related to the digging process and include the following:

- Nominal digging resistance,
- Nominal lateral resistance,
- Extraordinary digging resistance,
- Extraordinary lateral resistance,
- Support of the structure on the bucket wheel (bucket wheel excavators),
- Forces in the bucket chain (bucket chain excavators),
- Extraordinary support of the bucket wheel.

Figure 5.10 is a view of buckets with corner cutters (bucket-wheel excavator) and with teeth (bucket-chain excavator) and the schematic diagram of loads.

Figure 5.11 shows examples of results displayed as contours of equivalent stress.

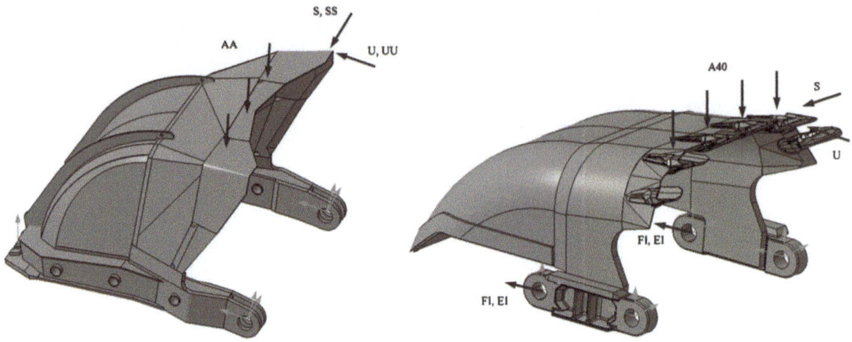

Fig. 5.10 Schematic diagram of digging forces acting on buckets with corner cutters (bucket-wheel excavator) and teeth (bucket-chain excavator)

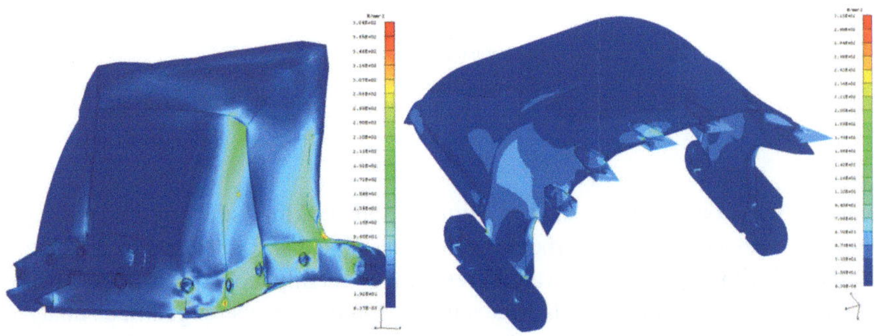

Fig. 5.11 Contours of equivalent stress according to Huber-Mises hypothesis (MPa)—buckets with corner cutters (bucket-wheel excavator) and teeth (bucket-chain excavator)

The above are examples of modernizations and optimizations of buckets, which are the first element of an excavating unit and which exhibit the highest rate of wear among all elements of the unit. Such projects are designed to improve the durability and service life of buckets and thus increase the uptime of machines. Of similar importance are factors related to energy consumption of the digging process. The presented approach to designing new buckets or modernizing existing ones enables full control of the decisive factors regarding the final optimum result.

5.2.2 *Load-Carrying Structure of the KWK jib*

As mentioned in the introduction, the necessity to introduce modernizations may stem from operational problems related to the dynamics of the machine. An example of such modernization is the modification of the jib structure for the bucket-wheel boom in the KWK 1500 excavator [12].

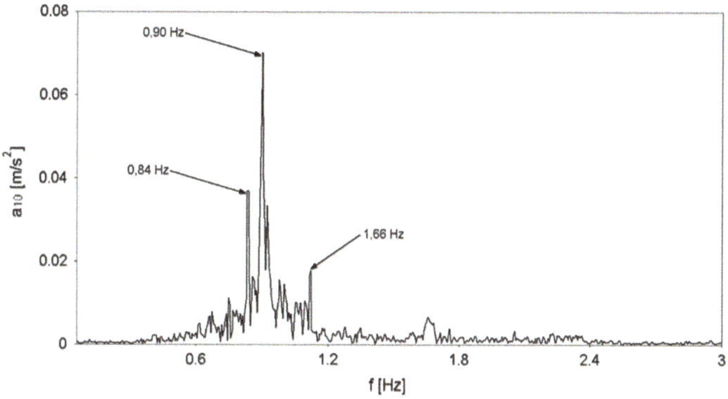

Fig. 5.12 Excavating towards the right before the modification—resonance frequency of 0.90 Hz

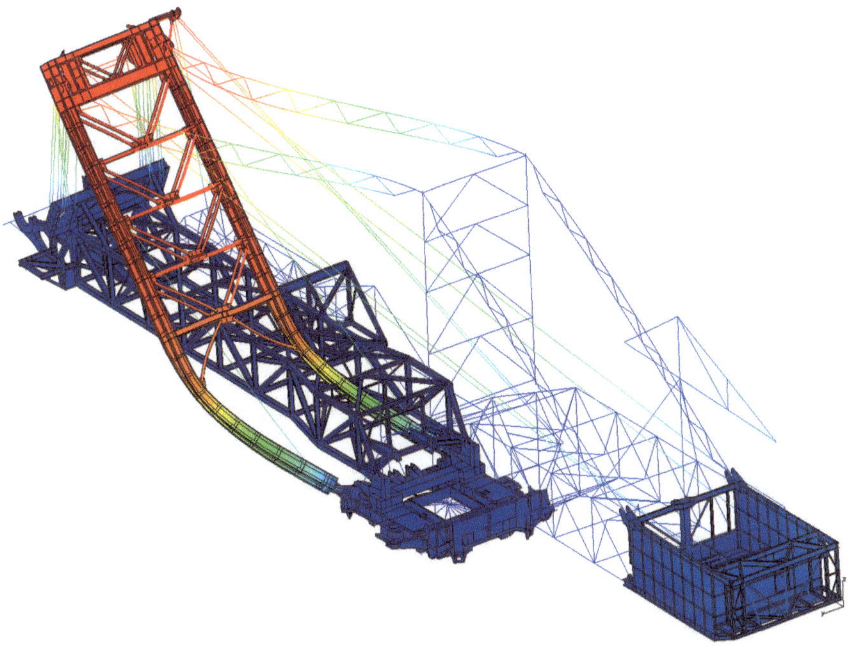

Fig. 5.13 Fifth mode shape of excavator superstructure—lateral vibration of the jib, $f = 1.07$ Hz

It was necessary to reinforce its lower section. Both the measurements and the computer simulations [13, 14] indicated that the lateral stiffness of the jib is too low. The identified frequency of lateral vibrations in the jib: 0.90 Hz in experimental tests (Fig. 5.12); and 1.07 Hz in numerical tests (Fig. 5.13) was in the frequency range of forced vibration, whose theoretical value is 0.87 Hz, which derives from the number of discharges from buckets: 52.15/min. This proximity of forced vibration frequency generated by excavating, and the frequency of jib vibrations, produces resonance vibrations with relatively high amplitudes.

An unwanted effect included high lateral vibrations of the jib during excavation and while traveling. In extreme cases, the jib hit the upper belt of the bucket-wheel boom. Since it is impossible to introduce additional stiffening lattices between the columns of the jib (collision with the bucket-wheel boom), the stiffeners were mounted on the exterior. The section of stiffeners was chosen so as to increase the lateral vibration frequency above the frequencies generated by the excavating process (number of discharges per minute). The assumed stiffness was one for which the frequency of lateral vibrations was 24% higher than the original value obtained in the FE model and 30% higher than the frequency measured on the machine. Additionally, strength analysis was performed for lateral loads. The objective of numerical analyses was to identify the state of stress in the stiffened section of the jib and the influence of this stiffening on the frequency of lateral vibrations in the jib. Figure 5.14 is a view of the discreet model with the introduced stiffeners of the jib.

Since it was difficult to calculate the lateral forces that occur during operation when the columns of the jib hit the upper girder of the bucket-wheel boom, the following assumptions were made. In the numerical model, simulations were carried out aimed at identifying the lateral acceleration that causes the jib to impact the

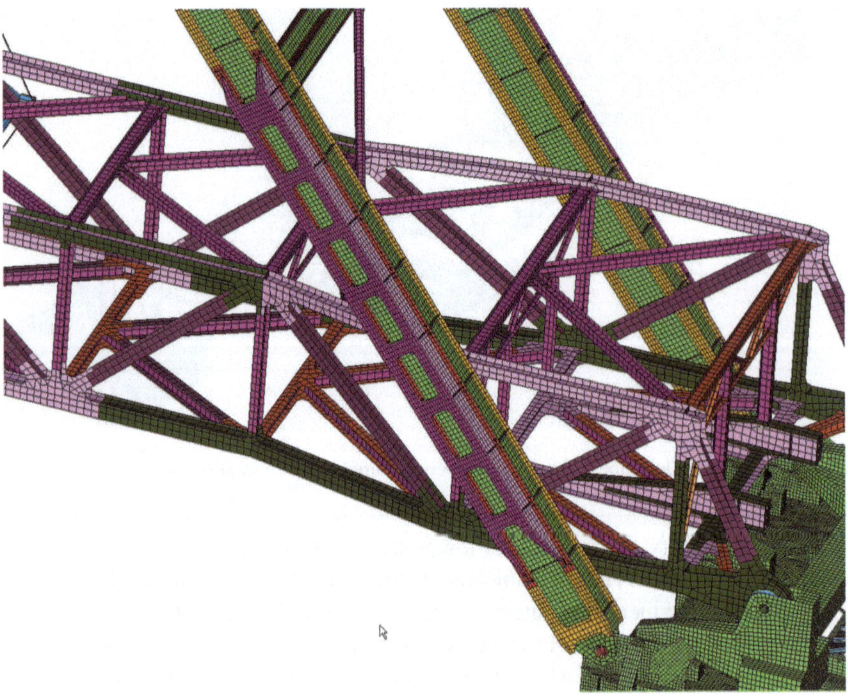

Fig. 5.14 Discreet shell model of the lower section of the jib with stiffeners

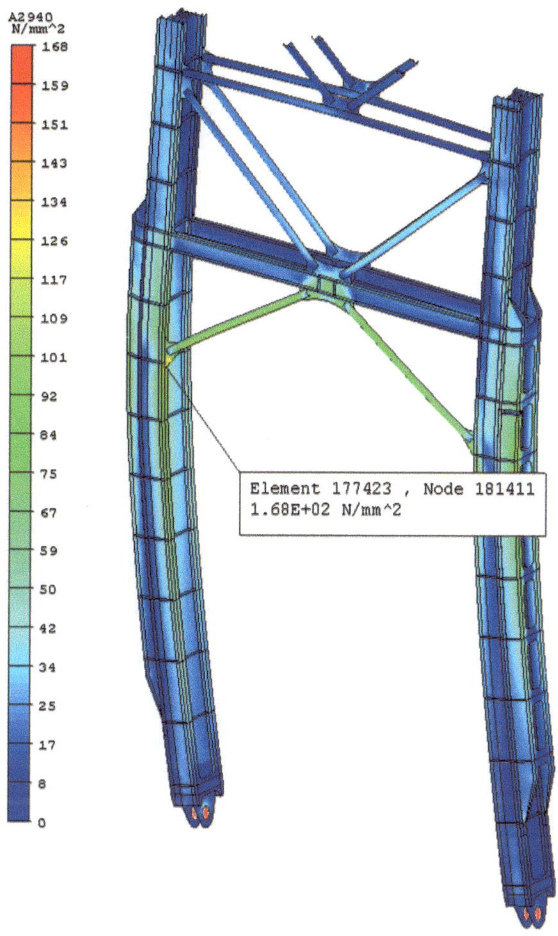

Fig. 5.15 Contours of equivalent stress in the jib under lateral load

upper girder of the bucket-wheel boom. The results of calculations indicate that such impact may occur if the value of forced acceleration is a = 2.94 m/s^2. In the case of the reinforced jib, the impact may occur only if the value of forced acceleration is a = 4.76 m/s^2. These values are theoretical and, in reality, the bucket-wheel boom also sways. Nevertheless, such considerations are needed for comparative analysis. Figure 5.15 shows the distribution of stress in the jib after introducing structural changes.

Figure 5.16 illustrates the mode shape of the jib after stiffening its lower section. The frequency of lateral vibrations in the jib, after introducing modifications based on FEM calculations, increased by approximately 24% and was equal to f = 1.21 Hz.

Numerical simulations served as the basis for designing structural modifications. The results of these modifications are shown in Fig. 5.17, which is a view of the

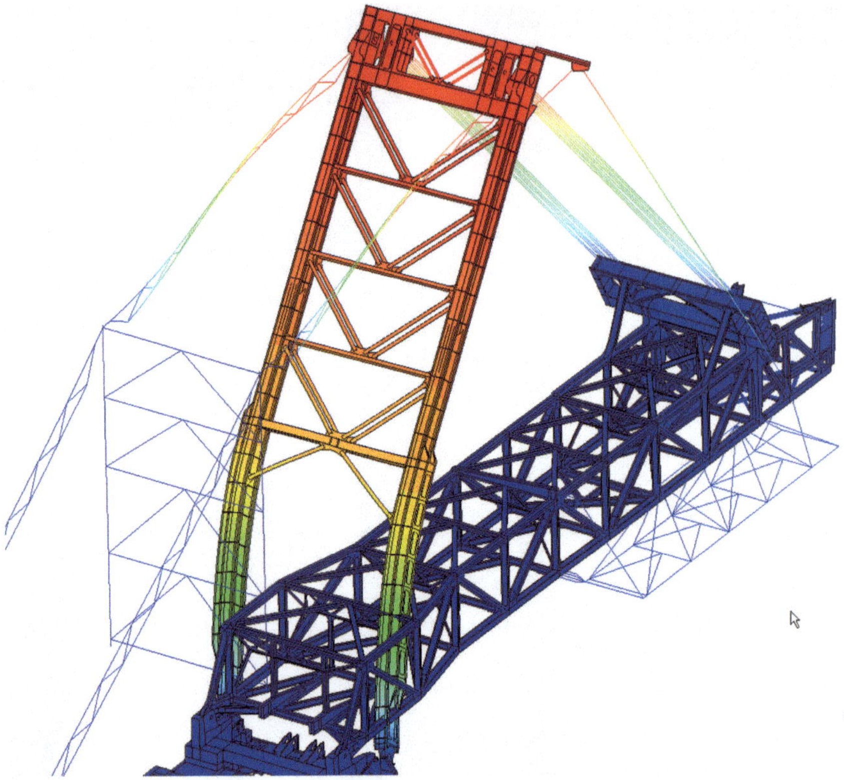

Fig. 5.16 Mode shape of the reinforced jib, $f = 1.21$ Hz

Fig. 5.17 View of jib with stiffeners

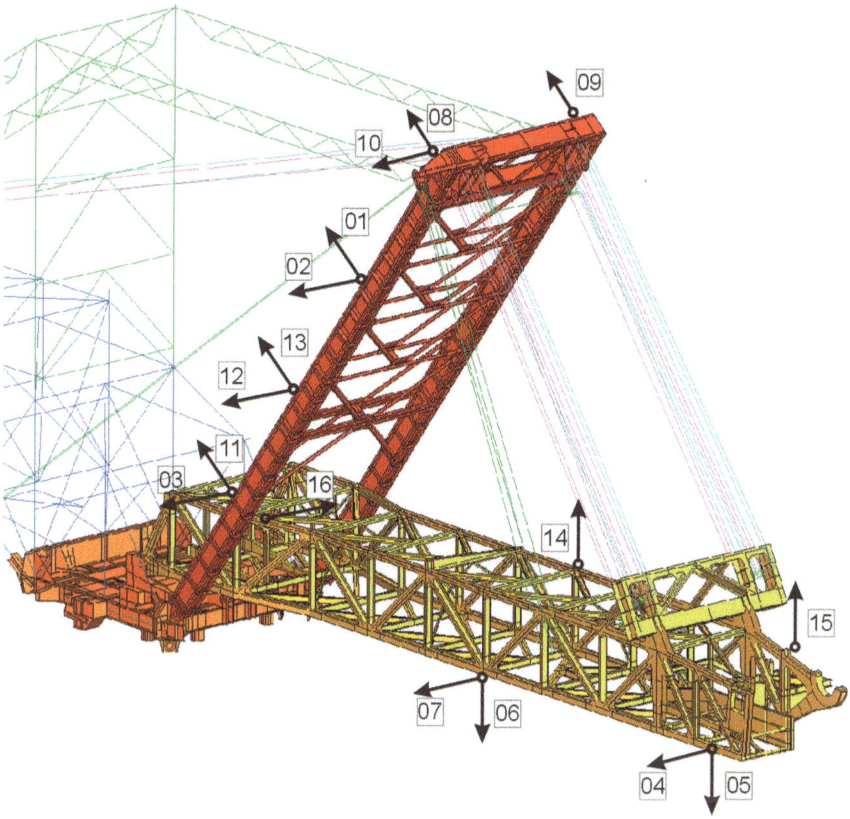

Fig. 5.18 Location of measurement points on the front section of the machine

reinforcements on the jib columns. The form of stiffeners blends well with the general structure of the jib.

After introducing reinforcements to the main columns (of the jib), frequency was measured in the same points as before modernization (Fig. 5.18).

The measuring system used was identical to the one used for measuring acceleration in selected points of the superstructure before modifications. Figures 5.19 and 5.20 show examples of frequency characteristics of vibrations measured at the measurement point located in the top section of the jib. The measured frequencies of lateral vibrations in the jib are approximately 20% higher than before modernization. The introduced modifications shifted the frequencies of structural vibration to higher ranges. As a result, the energy of the lateral mode shape of natural vibrations no longer causes excessive stress of the bucket-wheel jib structure.

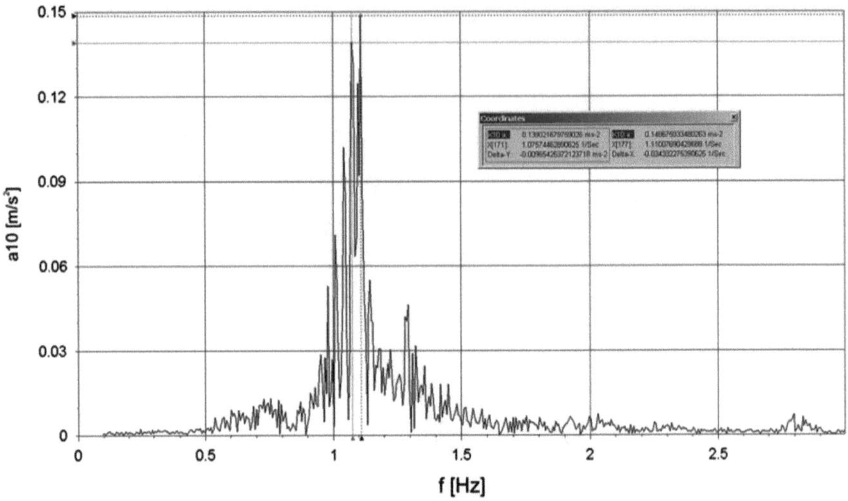

Fig. 5.19 Results of frequency analysis from the sensor in point no. 10, for measurements taken during excavation, $f = 1.11$ Hz

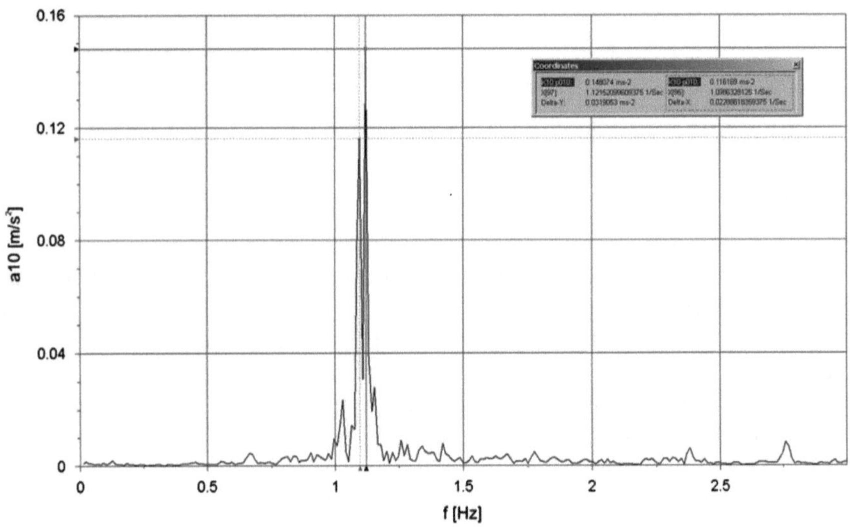

Fig. 5.20 Results of frequency analysis from the sensor in point no. 10, for measurements taken during travel, $f = 1.12$ Hz

5.2.3 *Modeling and Modification of Modal Characteristics of the KWK 1500.1 Excavator*

The object of numerical simulations is the new KWK bucket-wheel excavator with a bucket capacity of 1500 dm^3. The machine is primarily designed for excavating overburden and coal deposits in surface mines. The nominal frequency of discharges from buckets is 1.04 Hz. Standard [15] does not require calculations related to dynamics and modal analysis to be performed for load-carrying structures of surface mining machines. However, as shown by the previous example, problems with dynamics of lignite coal mining machines do occur and the aforementioned example is not an isolated case [16].

For the purpose of dynamic calculations, a complex beam and shell model was created [17, 18]. Numerical modal analysis was performed on the complete excavator superstructure. Elements of the undercarriage, such as the portal, crawler frame and steering frame were replaced by elastic elements with appropriate stiffness.

The objective of the simulation was to identify resonance areas. An analysis was carried out for a preliminary structural solution of the excavator superstructure. This was followed by a series of simulations in order to optimize the dynamic characteristics of the excavator.

The preliminary simulation demonstrated an overlapping of the harmonic frequency of forced vibration with the frequency of the sixth normal mode, which was equals to 2.03 Hz. The vibrations in this frequency range are caused by the low stiffness of the support system for the counterweight boom and the method of supporting the end of the boom on the central section of the excavator (Fig. 5.21).

The mode shape horizontal movement of the counterweight boom and vertical movement of its support system (Fig. 5.22).

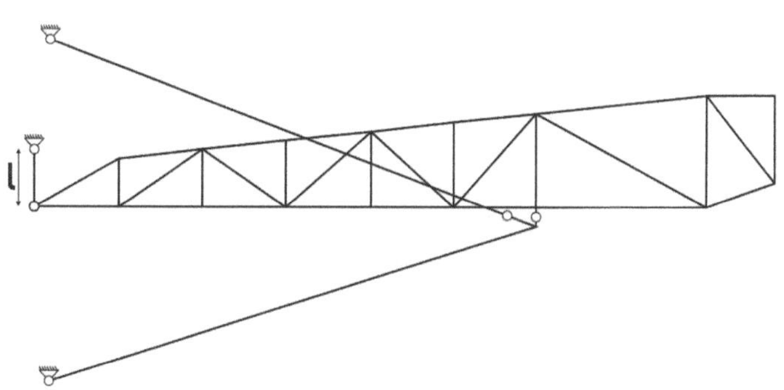

Fig. 5.21 Diagram showing the original support of the counterweight boom

Fig. 5.22 Sixth mode shape of natural frequencies

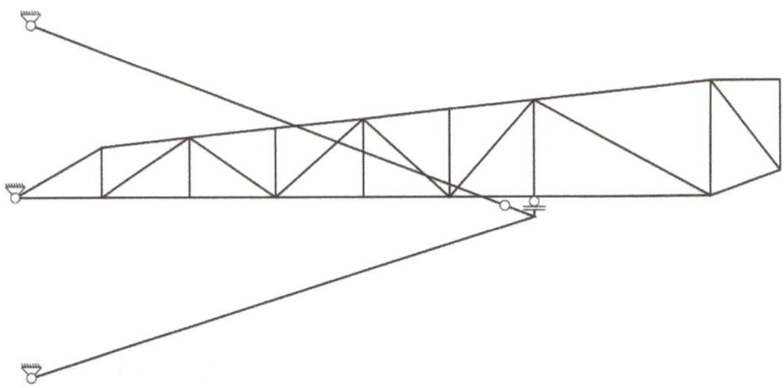

Fig. 5.23 Diagram of the final solution for supporting the counterweight boom

Simulations were performed to analyze several modification variants of the support and their impact on excavator dynamics. The length of the member "l" (Fig. 5.21) was modified from the original length of 2010 to 1050 mm and finally to 200 mm. The desired change in the frequency of the sixth mode shape was not observed. The introduced modifications did not have a significant influence on the other resonance frequencies.

Therefore, a change was made to the method of supporting the counterweight, which helped achieve the expected change in the dynamic characteristics. The member "l" was removed and only planar revolute joint was used. In the point where the counterweight is supported on tension members, a degree of freedom was added—movement along the horizontal longitudinal axis of the excavator (Fig. 5.23).

As a result, the resonance mode shape observed in the original variant of the excavator does not occur. However, another mode shape appeared with a much higher frequency (2.92 Hz). The deformation mode shape pertains mainly to the torsion of the counterweight boom and the bending of the bucket-wheel jib. This mode shape may cause the risk of resonance. The changes in the frequency of subsequent mode shapes for the analyzed structural solutions are presented in Table 5.1.

Table 5.1 Normal modes and frequencies

Mode shape	Frequency (Hz)				Mode shape
	l = 2010	l = 1050	l = 200	Solution 2	
1	0.671	0.679	0.696	0.708	1
2	0.829	–	–	–	–
3	0.835	0.803	0.802	0.827	2
–	–	0.838	0.860	0.894	3
4	1.65	1.62	1.62	1.64	4
5	1.86	1.86	1.87	1.88	5
6	**2.03**	**2.04**	**2.02**	–	–
7	2.47	2.50	2.50	2.49	6
–	–	–	–	2.92	7

The frequencies of the first three mode shapes increase with the limitation of counterweight displacement. There is a clear shift in frequency for the bending mode shape of the machine's booms in the horizontal plane. In the original solution, it was mode shape 2. When the member is shortened and the displacement of the boom is limited, this mode shape shifts to position 3. The introduced changes did not have a significant influence on the other mode shapes, except for the two aforementioned shapes.

Changing the modal characteristics of structures with a complicated, multibody construction (e.g., bucket-wheel excavators) is a complex process and requires a lot of experience. To obtain correct results, it is necessary to consider several joints between individual elements of the machine, including the complex hoisting system. Although the presented example is a new structure, it exhibits distinctive characteristics of modernization. This is due to the fact that the design of the KWK 1500.1 excavator was based on an existing structure of the KWK 1500 excavator. By modernizing old solutions in the KWK 1500 excavator, it was possible to determine and modify the dynamic characteristics of the new multibucket excavator at the stage of preliminary design. When the old type of supporting structure is changed, without change in the mobility of the system, significant changes are introduced to its dynamics. When the presented approach is applied to designing new machines that have a complex, multibody structure, the quality of design improves.

5.2.4 Modernization of Portals

Many years of research and observation of surface mining machines indicate that the damage that occurs in the portals of these machines is mostly caused by fatigue [19, 20]. The majority of this damage is located in the area of the slewing bearing. Examples of cracks are shown in Figs. 5.24 and 5.25.

Fig. 5.24 Examples of cracks in the portal of the SchRs 4600 × 30 excavator

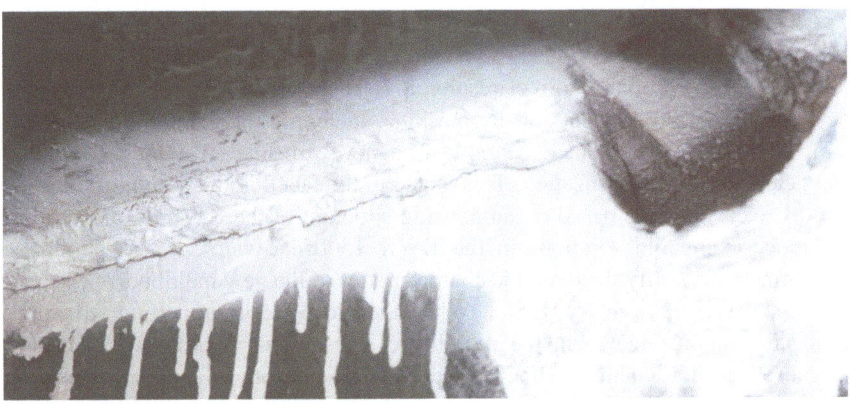

Fig. 5.25 Examples of cracks in the portal of the KWK 1500 s excavator

The appearance of such cracks may be related to faults in design, manufacture or operation. They can be divided into the following categories:

- Design faults—faulty or inaccurate calculation methods applied during dimensioning, insufficient stiffness of the portal frame in the undercarriage or its local excessive stiffness (sudden increase in stiffness), inappropriate choice of slewing bearing for the superstructure,

- Manufacture and assembly errors—poor quality of welded joints, and application of steel with strength parameters that are too low, with defects such as delamination,
- Operational faults—inappropriate balancing of the machine, excessive loads.

A case of damage and modification of the KWK 1500 s excavator portal is discussed below.

All of the KWK 1500 excavators operating in Polish mines have been found to sustain similar damage. Depending on the age of the machines (5–16 years old), the degree of degradation was diversified. Damage is located in the place where the cylindrical web of the girder connects with the top plate under the raceway of the turntable of the superstructure. In these machines, the rotation of the superstructure is performed by a swing mechanism equipped with four trolleys, whose wheels travel on the raceway attached to the portal. Such a method of supporting the superstructure results in large local loads on the wheels of the trolleys. Moreover, this system with four supports does not guarantee an even distribution of loads in all supports because the superstructure platform and portal are subject to uneven deformations resulting from the variable stiffness of their load-carrying structures.

In order to identify the causes of cracks, a surface geometric model and a shell discreet model of the portal were created. The geometric model of the portal is shown in Fig. 5.26.

The portal structure was calculated for the normal operation, load case H, which also allows for the structure to be tested with regard to fatigue strength. Experimental tests performed on the excavator demonstrated a shift of the axis of

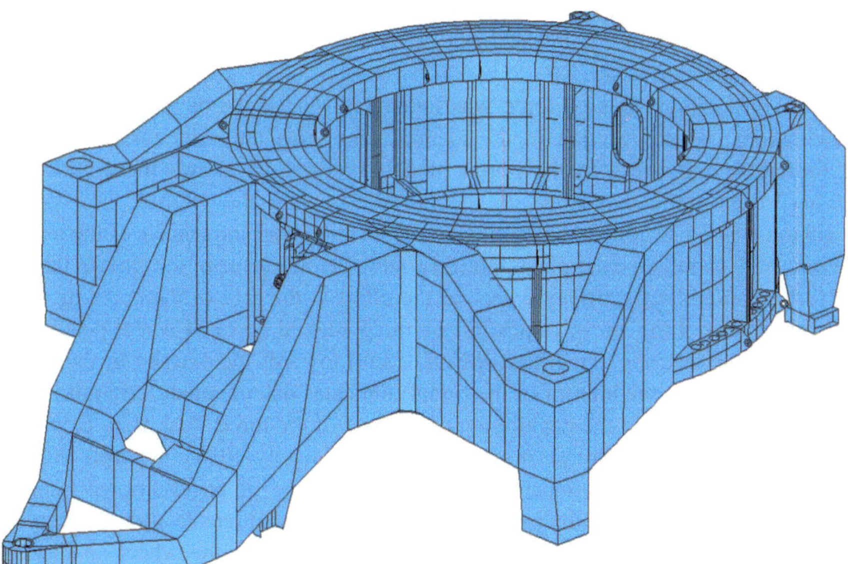

Fig. 5.26 Geometric model of the portal of KWK 1500 s excavator

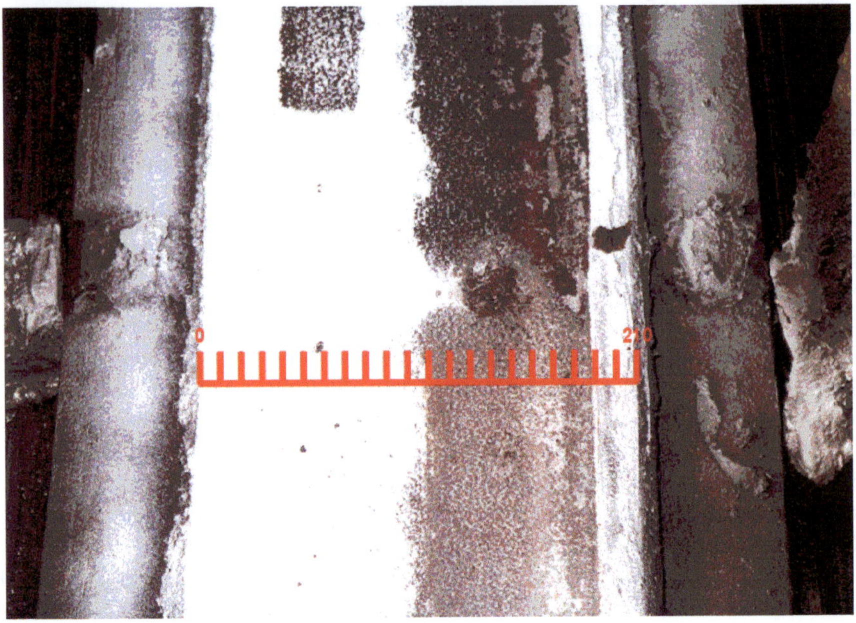

Fig. 5.27 Swing mechanism trolley wheel marks on the rail with an overlaid graduation scale in mm

forces exerted by the swing mechanism wheels in relation to the axis of the raceway of the swing mechanism. The maximum observed distance of the shift was approximately 50 mm. Figure 5.27 is a view of the wheel marks on the rail with an overlaid graduation scale.

The shift of the axis of force was taken into consideration during numerical calculations, which resulted in the identification of a three-dimensional area of deformation and stress in the portal model. Figure 5.28 shows the contours of Huber-Mises equivalent stress in the portal, showing the interior view of the cracking weld. The concentration of stress is visible under the wheels of the trolley. Figure 5.29 shows an example graph illustrating the influence of the shift in the axis of force exerted by the trolley wheel on the effort in the area of the cracking weld. A comparison of the stress values for the three studied cases of wheel positioning relative to the raceway axis (0: no shift; 50: shift towards the exterior of the portal; −50: shift towards the interior of the portal) indicates that the analyzed joint is very sensitive to changes in the positioning of the axis of the force. A shift by a distance of 50 mm causes an increase in stress by approximately 40% compared with the case without shift. This is caused by the appearance of an additional bending moment. The top plate of the portal bends on the web and tensile stresses occur on one side of the web, which add to the stresses exerted by the wheel on the rail,

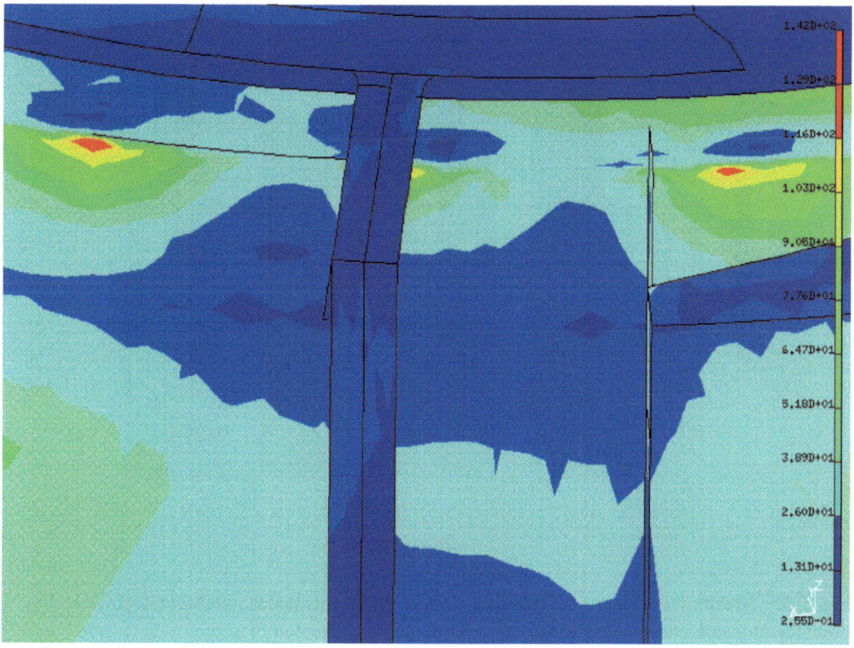

Fig. 5.28 Contours of equivalent stress in the portal structure in MPa, according to Huber-Mises hypothesis

whereas on the other side of the web, these stresses subtract. Apart from the cyclic repeated stress cycle, which results from the travelling of turntable wheels, the reversed stress cycle can also occur if there is a shift in the axis of forces exerted by individual wheels in opposite directions. A comparison between the stress range values obtained from calculations and the permissible values stated in the DIN 22261 standard [15] indicates that permissible stress is exceeded if there is even a minimum shift of the wheel axis in relation to the web of the portal.

Eliminating the causes of the shifting axis of wheel load in relation to the axis of the rail and web seems much more difficult than preventing the effects of this defect. Therefore, it was decided that reinforcements should be introduced to the top flange on the web in order to limit its bending, which would cause the loads from the swing mechanism wheel to distribute over a larger area and decrease the local deformations of the load-carrying structure of the girder. Moreover, such a reinforced flange should have additional supports, which would transfer the bending moment and, in the case of a cracked weld, connect the plate with the web.

Several variants of modifications were analyzed that could meet the aforementioned criterions. As a result, two equivalent structural solutions were obtained that decreased the stress level in the area of the damaged weld by approximately 50%.

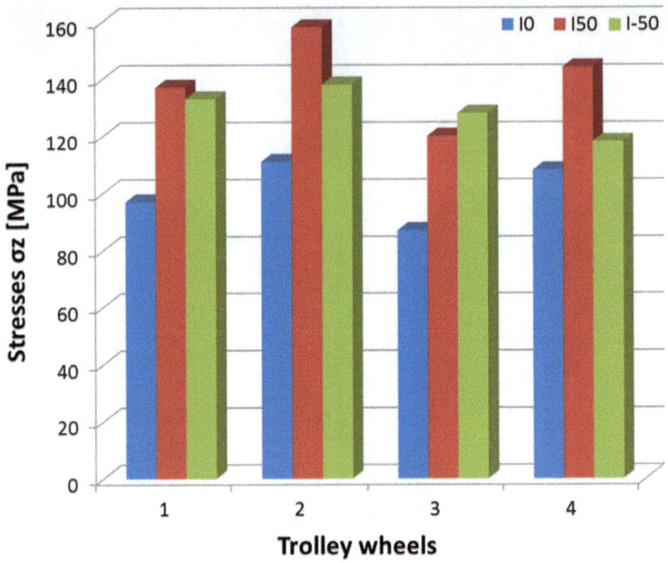

Fig. 5.29 Maximum σz stresses (z-vertical axis of the portal) under four wheels of the trolley for three locations of wheels on the raceway

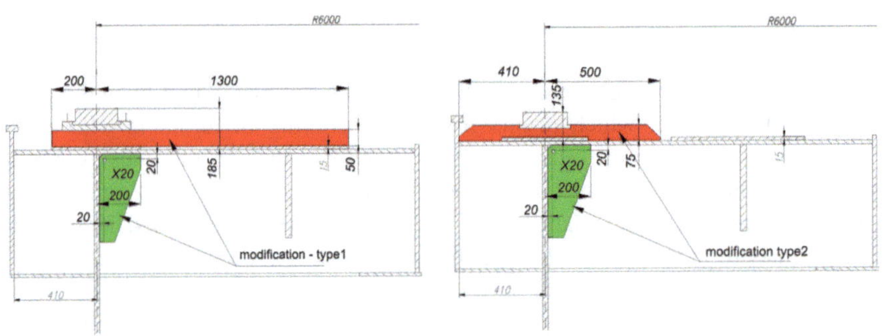

Fig. 5.30 Modifications of the platform

They effectively protected the structure against the unfavorable shift of the axis of load exerted by the wheel in relation to the axis of the raceway. Such a decrease in stresses was sufficient for the structure to meet the durability requirement in the DIN 22261-2 standard. By introducing these reinforcements, the strength of the joint also improved significantly. Eventually, two feasible solutions that offered similar results were selected. They are shown in Fig. 5.30. Figure 5.31 presents the contours of stress in the portal after introducing type two modifications.

Fig. 5.31 Contours of equivalent stress in the portal structure (interior view of the weld) in MPa according to Huber-Mises hypothesis—modification no. 2

The presented approach to modernization of elements of the portal may be successfully adopted for the whole load-carrying structure of the excavator or stacker. It eliminates the joints that are particularly stressed and thus hazardous to the entire machine.

Very often, modernizations also are made to elements of undercarriages of crawlers for surface mining machines [21–23]. Modernizations of selected machines are discussed in articles [24, 25].

Article [26] presents the scope of modernization works that have been carried out regularly and according to schedule since 1992 in the Polish Konin coal mine. The modernization works included mechanical electric and hydraulic elements and complete reconstructions and modernizations of load-carrying structures.

5.2.5 New Solutions Introduced to Surface Mining Machines

Along with the improvements and modernizations of surface mining machines, completely new solutions have also been introduced. A good example is the

Fig. 5.32 Vibroisolation system of the operator cabin

application of a modern system for vibroisolation of the operator cabin in a multibucket bucket-wheel excavator, which is presented in Fig. 5.32 [27].

Many different passive, semi-active and active damping systems are increasingly used to dampen the vibrations in large structures [28, 29]. New research is planned for the introduction of such damping systems to load-carrying structures of surface mining machines.

Other innovative solutions include the introduction of modern mechatronic systems to protect the excavation drive in the excavator as well as modern overload clutches [30]. The necessity to use such systems stems from the operating environment of bucket-wheel excavators. The excavated materials often contain rocks and/or other hard-mineable inclusions. The application of safety systems protects the load-carrying structure and machine elements against damage or destruction. The safety systems used to date have been purely mechanical. Their time of reaction was much longer and the possibilities of adjustment were limited compared with modern mechatronic systems. Such safety measures also prevent machine overload caused by unskilled operation.

Some of the aforementioned systems and many other modern solutions were implemented in the construction of the new Polish KWK 1500.1 excavator [31]. It is equipped with overload systems, monitoring systems, modern drives and other features. The load-carrying structure of this machine was designed and optimized using modern numerical methods in terms of ultimate, fatigue and buckling strength and with regard to the risk of unwanted vibrations. This structure is the culmination of many decades of experience, during which the modernizations of old machines were used to find the optimum solution for such structures. The KWK 1500.1 excavator is depicted in Fig. 5.33

Fig. 5.33 The newest excavator designed in Poland—KWK 1500.1

5.3 Case Study

This chapter focuses on the modernization of the excavating unit in the SchRs 4600 excavator. A separate section is devoted to this modernization because of the comprehensive approach and consideration of many factors. As described in the following section, choosing a new excavating unit is a rather complex task. More importantly, the numerical and experimental method presented below has not yet been applied in the construction of excavating units of bucket-wheel excavators. The excavating unit should be understood as buckets, teeth, bucket wheel and drives.

The SchRs 4600.50 (Fig. 5.34a) and SchRs 4600.30 (Fig. 5.34b) bucket-wheel excavators operate in a lignite surface mine. Although the two machines differ in class [12] and size, they have the same excavating unit. The first machine, without the discharge bridge, weighs approximately 5000 Mg. It is over 60 m in height and over 120 m in length. The second machine weighs approximately 3300 Mg, has a height of over 36 m and a length of over 70 m. In both machines, the excavating unit is powered by three engines, each with 530 kW of power. The assumptions regarding the excavating unit do not fit well with Polish geological conditions. The 11 buckets mounted to the wheel caused strong dynamic impulse load resulting in numerous damages to the superstructure and drives. The current solutions also do not guarantee sufficient dredging. In order to minimize the unfavorable effects in the SchRs 4600.30 excavator, 11 additional semi-coated buckets for coal excavating

Fig. 5.34 **a** SchRs 4600.50 excavator; and **b** SchRs 4600.30 excavator

(Fig. 5.34b) were mounted. This improved the operating conditions of the excavator, especially the coal dredging.

The abovementioned solution to operational problems was only temporary. For the final solution, a decision was made to replace the bucket wheel with one that had more buckets. The main prerequisite for modernization was to install the same bucket wheel and buckets in both SchRs 4600 excavator types. This prerequisite imposes the necessity to conduct detailed analysis of dynamic characteristics for both machines in order to establish the optimum number of buckets. This means that the frequency of forced vibrations determined by discharges from the new wheel must not overlap with any of the natural frequencies of both machines. Additional criteria concerned the efficiency of the machine and the necessity to adapt it to the existing drives. Previously, in the system with 11 buckets, the frequency of forced vibrations was approx. 0.57 Hz. In order to precisely determine the dynamic characteristics of the machines a comprehensive study was conducted using the experimental and operational modal analysis [32, 33].

The discrete model of the complete superstructure of the machine consists of a slewing platform, counterweight boom, central part, bucket-wheel boom and front and rear tower. The numerical model did not include undercarriage elements, such as the platform, elements of the driving system or the discharge bridge. Those elements were not taken under consideration because the most problematic source of vibrations is the superstructure. However, the stiffness of the undercarriage elements in the model must be compensated for in a different manner because they influence the global modes of the machine. In the described discrete model, the superstructure of the SchRs 4600.50 bucket-wheel excavator was supported in the area where the raceway of the main bearing is located. The stiffness of the reduced elements was adjusted to the stiffness of all the undercarriage parts. This approach provides only an approximated representation of the real conditions. However, assuming that the whole undercarriage was included in the simulations, the ground stiffness is still unknown and even a perfect model of the whole machine would not produce the exact value of stiffness. The final conclusions indicate that the simplified approach offers satisfactory results. Figure 5.35 shows the first mode shapes

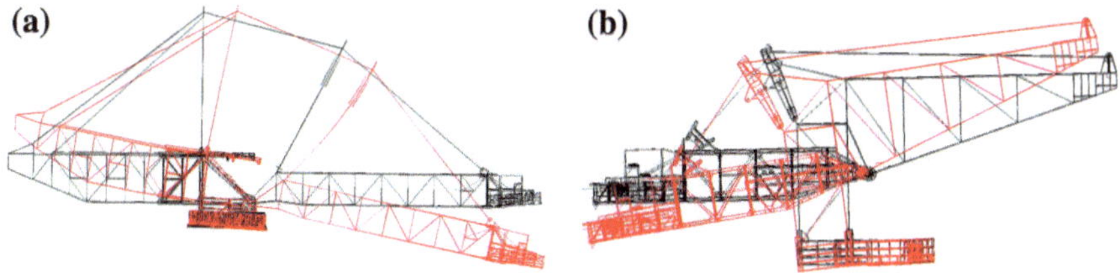

Fig. 5.35 **a** First mode shape of the SchRs 4600.50 excavator; and **b** the SchRs 4600.30 excavator

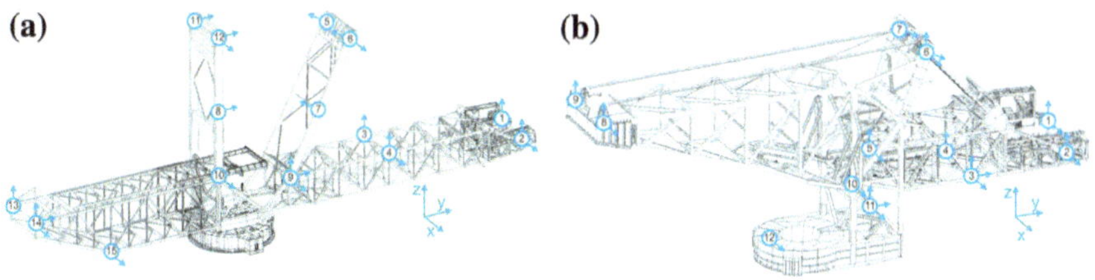

Fig. 5.36 **a** Measurement points: SchRs 4600.50 excavator and **b** SchRs 4600.30 excavator

in superstructures of both excavators, whose frequencies are 0.28 and 0.40 Hz for the SchRs 4600.50 and the SchRs 4600.30 excavators, respectively.

Numerical measurements served as the basis for the proper modal experiment in real conditions. First, the measurement points were selected (Fig. 5.36).

Measurements were taken under various operational conditions that can be observed during the lifetime of the excavator. From all the data sets, three of the most reliable sets were selected for analysis: upward excavation, downward excavation and machine travel.

Regardless of the load case, no significant changes were observed in the correlation between the numerical and experimental models. Attempts were made to increase the correlation between the numerical and experimental modal models.

When comparing both models, it is important to investigate not only the similarity of the mode shapes, but also its relation to the frequency of the compared modes. Therefore, in order to facilitate this, the author's generalized correlation factor (k_u) was used in the following form (2):

$$k_u(f_b) = \mathrm{MAC_{ab}}\left(1 - \left|\frac{f_a - f_b}{f_b}\right|\right) \tag{5.2}$$

where f_a is the frequency of the mode from the numerical model, f_b is the frequency of the mode from the experimental model, and $\mathrm{MAC_{ab}}$ is the correlation level of the numerical (a) and experimental (b) mode.

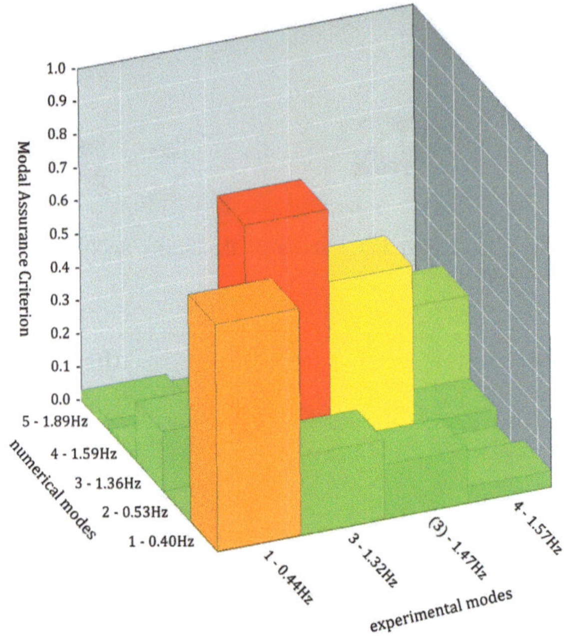

Fig. 5.37 MAC matrix for the SchRs 4600.30 excavator

The modifications introduced to the numerical model include boundary conditions resulting from the changes in operational conditions. Figure 5.37 presents an example MAC matrix for the numerical and experimental models of the SchRs 4600.30 excavator.

Based on numerical and experimental modal characteristics, it was possible to define such a number of buckets so that the vibration generated in the excavating process did not correspond to any of the resonance frequencies. Of equal importance was the influence of change in the mass of the new bucket wheel and buckets on the identified modal characteristics. The first solution included 17 buckets [34]. Eventually, however, 18 buckets were mounted on the wheel. This was due to the fact that the newly designed wheel was heavier than initially planned. The change in mass influenced the shift of resonance frequencies. As a result, it was also necessary to increase the frequency of forced vibration.

The new bucket-wheel design is shown in Fig. 5.38.

The bucket wheel was also equipped with new buckets with a design that took into account modern solutions discussed in Sect. 5.2. The structural solution for the new buckets is shown in Fig. 5.39.

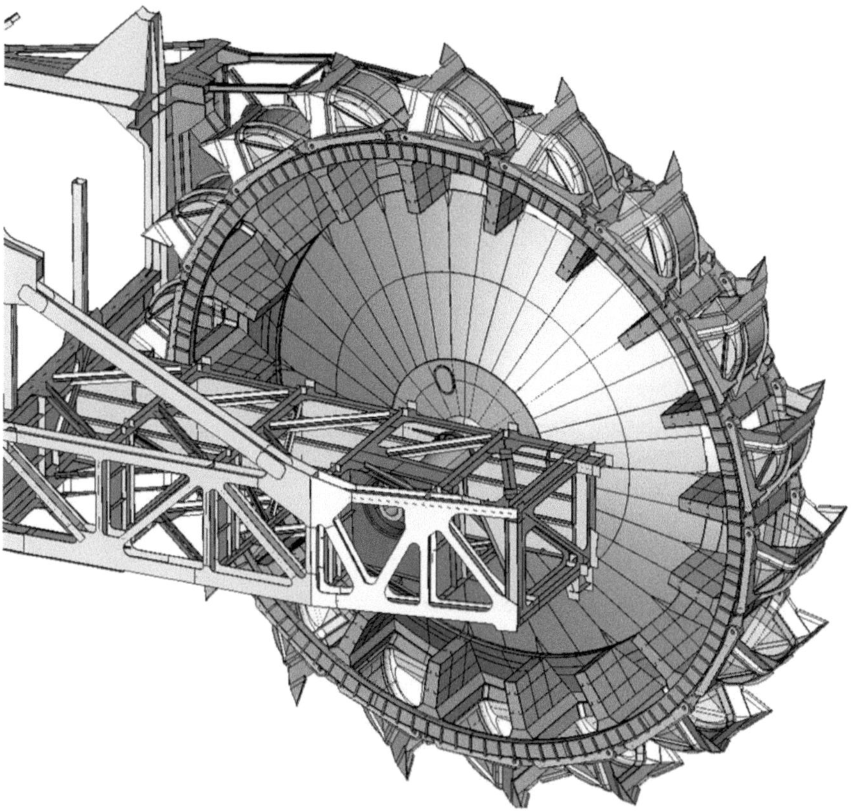

Fig. 5.38 New bucket wheel with 18 buckets

The basic parameter of the number of buckets was determined by the resonance characteristics of the machine, which were identified using numerical and experimental methods. This approach has not been used before and is an innovation in the design of surface mining machines. Choosing the number of buckets was additionally complicated by the fact that the generated forced vibration, which depended on the number of buckets, had to be adjusted to the resonance characteristics of three excavators simultaneously (one SchRs 4600.50 excavator, two SchRs 4600.30 excavators).

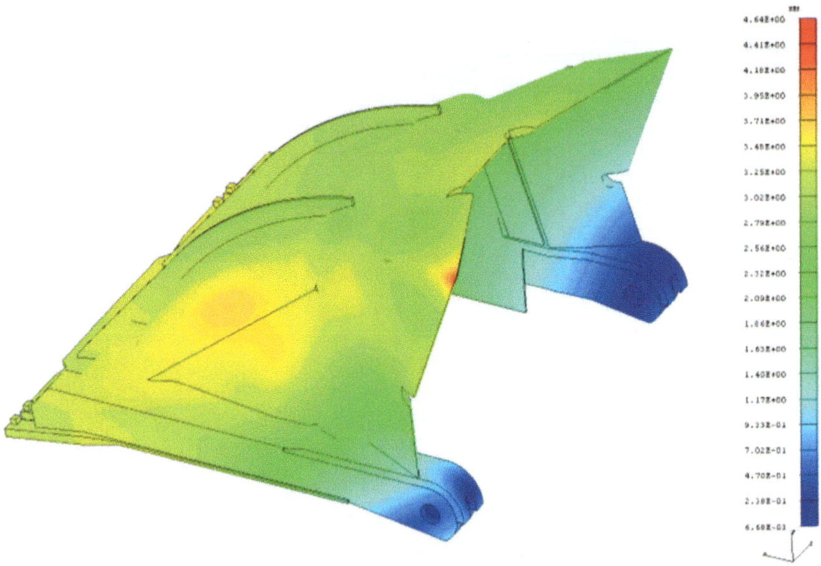

Fig. 5.39 Calculation model of the new bucket

The final solution with 18 buckets generates a frequency of forced vibration equal to 0.93 Hz. The rotation speed of the bucket wheel was not changed and is equal to 3.11 rpm, which gives 56 discharges per minute. In order to obtain the theoretical productivity of the excavator at the level of 9337 m^3/h, the capacity of the new bucket is 2139 L.

References

1. Rusiński E, Moczko P (2011) Modernization of the excavation unit of excavators SchRs 4600 Górnictwo i Geoinżynieria, R. 35, vol 3/1, pp 217–230
2. Rusiński E, Smolnicki T, Moczko P (2005) Modification of SchRs 4600 × 30 excavator body's slewing bearing and its supporting structure. World Min 57(3):2–8
3. Araujo LS, de Almeida LH, Batista EM (2009) Analysis of a bucket wheel stacker reclaimer structural failure. In: Proceedings of the conference "materials science & technology 2009", Charles R. Morin Memorial symposium on failure analysis and prevention. Proceedings on CD, Pittsburgh, http://www.matscitech.org
4. Savkovic M, Gašic M, Arsic M, Petrovic R (2011) Analysis of the axle fracture of the bucket wheel excavator. Eng Fail Anal 18:433–441
5. Arsic M, Bošnjak S, Zrnic N, Sedmak A, Gnjatovic N (2010) Bucket wheel failure caused by residual stresses in welded joints. Eng Fail Anal
6. Hawrylak H, Sobolski R (1967) Maszyny podstawowe górnictwa odkrywkowego. Wydawnictwo Śląsk, Katowice
7. Kasztelewicz Z (2012) Bucket wheel excavators and spreaders. Technology of operation, Art-Tekst, Kraków

8. Pieczonka K (2007) Engineering of earth moving machines. Part 1. Basics of excavation, travel, hoisting and slewing. Wroclaw University of Technology Publishing House
9. Raaz V (1999) Assessment of the digging force and optimum selection of the mechanical and operational parameters of bucket wheel excavators for mining of overburden, coal and partings. Krupp Fördertechnik, Esen
10. Rusiński E, Czmochowski J, Smolnicki T (2000) Advanced finite element method in load carrying structures. Wroclaw University of Technology Publishing House
11. Zienkiewicz OC, Taylor RL (1990) The finite element method, 4th edn. McGRAW-HILL BOOK COMPANY, London
12. Czmochowski J (2008) Identification of modal models of excavators in lignite mining. Wroclaw University of Technology Publishing House
13. Pietrusiak D, Czmochowski J (2011) Bucket wheel excavator dynamics investigation. Selected problems of modal analysis of mechanical systems. In: Uhl T (ed) Wydawnictwo Naukowe Instytutu Technologii Eksploatacji – PIB, pp 109–117
14. Pietrusiak D, Rusiński E, Smolnicki T, Kowalczyk M (2011) Dynamic effects in spreader operation. Transport przemysłowy i maszyny robocze 1(11):66–68
15. DIN 22261-2 Bagger, Absetzer und Zusatzgeräte in Braunkohlentagebauen
16. Babiarz S, Dudek D (2007) Annals of failures and catastrophes of surface mining machines in Poland. Wroclaw University of Technology Publishing House
17. Bosnjak S, Zrnic N, Oguamanam D (2006) On the dynamic modeling of bucket wheel excavators. FME Trans 34:221–226
18. Rusiński E, Czmochowski J, Pietrusiak D (2012) Problems of steel construction modal models indentification. Eksploatacja i Niezawodność – Maintenance and Reliability 14(1):54–61
19. Rusiński E, Moczko P (2002) Cause of the damage of the chassis girders of the bucket wheel excavator. Przegląd Mechaniczny 4'02
20. Rusiński E, Smolnicki T, Moczko P (2005) Modification of SchRs4600 × 30 excavator body's slewing bearing and its supporting structure. World Min—Surf Undergr 57(3)
21. Bosnjak S, Gnjatovic N, Zrnic N (2010) Redesign of the connecting eye-plate of the bucket wheel boom hoisting system. In: Machine design—50th anniversary of the faculty of technical science, May 18th, 2010, Novi Sad, Serbia
22. Bosnjak S, Petkovic Z, Gnjatovic N, Milenovic I, Jerman B (2013) Impact of the track wheel axles on the strength of the bucket wheel excavator two-wheel boogie. Tehnički vjesnik 20(5):803–810
23. Wocka N, Warcholak A (2008) Crawlers of the surface mining machines—design solutions for durability and reliability. Węgiel Brunatny (3):64
24. Kośla W (2007) Modernizacja koparki KWK-1500 s I w Kopalni "Adamów", Węgiel Brunatny (1):58
25. Paszkowiak J, Wojnicz M, Woźniak Z (2006) Modernization of the excavator Rs-400—hydraulics. Węgiel Brunatny (1):54
26. Paszkowiak J, Maniak M, Wonicz M, Woźniak Z (2013) Modernizations of surface mining machines—unique experience in Konin Mine. Węgiel Brunatny (3):84
27. Rusiński E, Lewandowski T, Słomski W (2004) Vibration isolation systems of wheel excavators operator's cabins. Konstruktion Juli/August 7/8-2004
28. Weber F, Feltrin G, Maślanka M, Fobo W, Distl H (2009) Design of viscous dampers targeting multiple cable modes. Eng Struct 31:2797–2800
29. Weber F, Maślanka M (2014) Precise stiffness and damping emulation with MR dampers and its application to semi-active tuned mass dampers of Wolgograd Bridge. Smart Mater Struct 23:015019 (18 pp)
30. Kowalczyk M, Rusiński E (2013) Adjustment of the characteristics of the overload couplings in the main drives units of bucket wheel excavators, Elbląg rack and pinion gears: theory and practice. In: Rybczyńska T, Łabanowski J (eds.). Elbląg, pp 15–31

31. Majkowski A, Bielecki B, Kowalczyk M, Krasucki R, Moczko P, Olejarz J, Rusiński E, Stassek J (2012) Bucket wheel excavator KWK-1500.1—a new machine in KWB Turów Mine, Węgiel Brunatny (4):81
32. Heylen W, Lammens S, Sas P (2007) Modal analysis theory and testing. Katholieke Universiteit Leuven
33. Rusinski E, Czmochowski J (2001) Die Modalanalyse des Oberbaus eines Baggers vom Typ SchRs-800. Surf Min Braunkohle Other Miner 53(3):319–324
34. Rusinski E, Dragan S, Moczko P, Pietrusiak D (2012) Implementation of experimental method of determining modal characteristics of surface mining machinery in the modernization of the excavating unit. Arch Civil Mech Eng 12(4):471–476

Chapter 6
Summary and Conclusions

This book is a collection of practical information on the operation, maintenance requirements and modernization of the largest mobile machines built by man, namely surface mining machines: bucket-wheel and bucket-chain excavators and stackers [1]. The abovementioned aspects are not yet fully understood due to many factors, such as: the size of the machines, which necessitates unique technical solutions, especially for their load-carrying structures; the fact that only single units are produced for long-term operation, which often extends beyond their lifetime; and, finally, the variable operating conditions of surface mining machines, particularly excavators, which are difficult to predict at the designing stage.

This situation leads to operational problems associated with various types of degradation processes that affect these machines after long-term operation. Because of their prototypical nature, it is difficult, bordering on impossible, to draw useful conclusions regarding their operation that could then be applied to other machines, or to determine the optimal maintenance policies. Because of the abovementioned factors, using surface mining machines is still a major challenge that often requires special solutions in order to provide maximum operational safety. Of particular importance in this context is the experience gained through many years of operation and maintenance of the mentioned machines. This experience serves as a valuable source of information that can be used to support decisions about both short-term and long-term modernization schedules, technical maintenance policies or removing dangerous machines from service. Such a know-how, combined with the latest achievements related to technical condition assessment methods, residual life prediction, and implementation of innovative technical solutions, allows a dynamic development of new surface mining machines and effective and economic modernization of existing ones.

The authors of this book see the need to gather and systematize information related to the operation and modernization of surface mining machines. This is because of the absence of such publications in the literature. The little information that exists in the form of books or standards is mostly outdated, given the current state of science and technology. This book presents problems related to the

E. Rusiński et al., *Surface Mining Machines*,
DOI 10.1007/978-3-319-47792-3_6

operation of surface mining machines, which are based on current knowledge and experience gained over many years of research on such machines. The book also discusses the causes, consequences and possible methods of preventing such problems in the future. Also described is a unique approach to technical condition assessment of such machines, which is based on numerical and experimental methods. In addition, the book identifies optimal strategies for technical maintenance and presents methods and directions for modernizing existing machines and for developing new machines. All of these aspects are supported by examples of activities involved in technical condition management of machines currently in operation and in the design of new machines [2].

The first chapter addresses issues related to the operation of surface mining machines. A general description is included of the multi-layered research problem discussed in this book. This explains the importance of the issues in this book and emphasizes its current significance.

What follows is a systematization of the structural forms of surface mining machines in terms of their functions and their size, which necessitates diversified technical solutions. Next, the operational problems of surface mining machines are presented and grouped according to their functions carried out in the individual stages of the mining process, in the case of excavators, and the stacking process, in the case of stackers. Also described are the main mechanical and structural subassemblies of bucket-wheel excavators and stackers in which operational problems occur. The focus is on those structural and mechanical components that are prone to degradation processes and that affect the machine's uptime and operational safety. Relevant examples of degradation-related events are also included.

The phenomena that occur during excavation and stacking (e.g., digging forces, dynamic phenomena, loads in kinematic nodes, traveling) constitute an important part of this chapter. Their impact on individual components of the machine can be evaluated and thus their significance in the processes of shear and fatigue damage can be determined. Loads resulting from these phenomena are grouped according to the type of interaction. Such classification facilitates the decision-making process in matters related to technical maintenance and directions of future modernizations.

Subsequent sections discuss the disadvantages of traditional methods of dimensioning surface mining machines, which are based mainly on the relevant standards and use analytical calculations. It has been proven that this process is incompatible with current technical and operational requirements for surface mining machines.

The third chapter describes the problems of technical condition assessment of surface mining machines. A properly conducted technical condition assessment is required to use mines in a reliable and economically justified manner. The book demonstrates that the diagnostics of such machines is not particularly popular, despite the undeniable benefits of obtaining information on their technical condition, which changes over time thereby requiring that operational decisions reflect the current condition. This state of affairs is mainly due to the lack of knowledge and experience of operators and maintenance engineers. The current state of technology allows for continuous or periodic condition monitoring of such machines. The following three basic strategies of machine operation may be applied for this purpose:

- Operation until failure occurs (corrective maintenance),
- Condition-based operation (preventive maintenance),
- Operation that prevents and correct failures (predictive maintenance).

It has been determined that approximately 90% of surface mining machines are operated using the non-optimal first and second strategy, of which the first operation strategy, i.e., until failure occurs, is the least advantageous [3]. Technical condition assessment in this case amounts to periodic inspections and identification of the causes of faults, should they occur. This approach often leads to unplanned downtime, which significantly reduces the efficiency of production.

Given the aforementioned state of technical condition maintenance, and on the basis of over 30 years of experience designing, researching and modernizing surface mining machines, the authors have developed a comprehensive method of assessing the technical condition of such machines, which is described in detail in chapter three. This method can predict the operation of machines with consideration of their technical condition as well as current and future operating conditions. Additionally, the chapter describes the main components of the method (experimental measurement methods and techniques and calculations), and the entire methodology that can be used in two modes depending on the required results and methods of predicting residual life. The first mode uses a modified classic method of selecting the characteristics of a structural node with a detailed geometric description by means of 3D numerical models. The second mode analyzes in detail the lifetime of structural nodes taking into account their actual geometry, location of joints (e.g., welded, bolted joints) and spectra of real loads that cause cumulative fatigue damage. The first approach provides information on the possibility of failure in a specific location, while the second mode additionally specifies when such damage may occur and what the potential residual life is until failure occurs. Both variations of the method are illustrated by examples.

The next chapter is the study of failures and disasters, which are an inseparable element of machine operation, especially in extreme conditions, such as in the case of mining. This chapter demonstrates that damage is largely correlated with operating loads that exceed the design requirements. However, it is not the design that is faulty, since the design follows the requirements set in the standard. As shown in this book, the problem lies in the incompatibility of these requirements with the real phenomena, mainly dynamic in nature, which occur during the operation of surface mining machines in difficult and diverse (heterogeneous) geological conditions [4]. There are situations when such phenomena are caused by faulty operation of the machine. There is also a category of causes of damage that consists of various material and manufacturing defects.

To prevent damage or destruction of the machine or its assemblies, it is necessary to monitor the status of its key elements and take corrective actions. Properly scheduled repairs can be crucial in this case. Unfortunately, the construction of monitoring systems and the proper observation/monitoring of all elements in surface mining machines is often extremely difficult, or even impossible, due to their complexity or for financial reasons. As a result, there have been cases of disasters that could have been avoided (i.e., symptoms were visible to the naked eye), but where it was not possible to monitor the component that caused the damage as it was difficult to reach. It is precisely in such cases when the new method of technical condition assessment described in this book can be used to identify areas that are critical in terms of their technical condition and safety of the machine, and which are designated for periodic inspections at specified time intervals. With the development of new technologies and construction of new machines, continuous condition monitoring has been introduced for critical areas in the structure, which provides much more information in comparison with periodic inspection and helps to develop a better strategy of operation. This chapter also presents a range of activities that are carried out after the damage is revealed and whose purpose is to find the cause of such damage. Combined, these activities form a complete methodology, which for years has been successfully used and developed by the authors of this book.

The fifth chapter describes the most interesting modernization projects that have been carried out in recent years and in which the authors of this book have participated. These include both minor improvements (e.g., attaching a bucket) as well as complete modifications to key components (excavating unit of SchRs 4600 excavators). These projects also serve as evidence of the numerous imperfections in machines that are currently in operation. It should be noted that although new designs are based on existing solutions, these machines are still prototypical in nature due to the complexity of their entire structure. In addition, as already mentioned, because of their long-term operation, which typically extends beyond their designed life, new problems/damage constantly emerge, thus necessitating further changes. The most challenging components to modernize are steel load-carrying structures. They serve as the skeleton for all other subassemblies such as: machine assemblies, electrical assemblies, communication, and so on. The modernization/replacement of these subassemblies is not very difficult as they only have a local effect on the whole structure. However, in the case of the steel load-carrying structure, which transfers the main loads and is responsible for machine stability, it is usually difficult to repair/replace it or to modernize its individual components. In such projects, the load-carrying function of the entire machine must be retained or replaced by another, temporary load-carrying structure, which is expensive, time-consuming and requires complicated assembly procedures. Such projects do get carried out, however, in order to ensure safety of the machines. Apart from purely technical drivers for changes resulting from the machine itself, there are also external factors—geological, ecological/political or economic. Currently, apart from the desire to extend the operating time, the main aim is to reduce the costs of mining. This involves a decrease in energy consumption and an increase in the uptime of machines operating in mines.

To summarize, the book covers the following:

- It identifies the problems related to the operation, maintenance and modernization of heavy performance machines,
- It offers practical solutions to the presented problems (case studies),
- It presents research that is of interest to mechanical engineers,
- It describes unique (large scale) machines,
- It discusses the implementation of known solutions/methods and proposes a new approach to the identified problems.

The structures described in this book, i.e., surface mining machines, require constant development, in the case of new machines; and modernization, in the case of existing machines. Due to the complexity of the phenomena resulting from their size and from the variability of their operating conditions, these tasks cannot be carried out in a typical manner. Modernizations of existing structures constitute the greatest challenge. In this regard, major modifications (e.g., replacement of the entire body of the machine) are rarely made. In contrast, local improvements such as the modernization of the excavating unit, installation of protection systems against overloads, local structural changes to load-carrying structures, are the main focus of activities aimed at maintaining the required technical condition or adapting it to current technical and operating conditions. An additional trend can be seen with the shift towards effective technical condition assessment and monitoring solutions, which are evolving from the periodic approach to the use of continuous surveillance systems. Monitoring signals, for which significant change is observable only over a long time horizon, constitutes a challenge in this case but the current state of knowledge and technology can solve such problems [5].

This work is addressed to the following groups of potential readers and for the following reasons:

- For those interested in engineering, the problems of mega-machines will surely be fascinating.
- For academics, this new publication will be useful in education and research.
- For the industry, this publication will serve as a handbook on the operation, maintenance and modernization of machines. It offers solutions to the identified problems and new ideas for modernization and maintenance planning. This in turn can lower the costs of operation, prevent catastrophic failures, reduce downtime and help to increase the durability of machines.
- Designers of high performance machines and large scale steel structures will learn about the common problems that can occur during operation and will have a chance to solve some of them in the designing stage.

The authors hope that this work and the problems it addresses will contribute to the development of knowledge and lead to more informed and, above all, purposeful and reasonable actions aimed at improving the safety and efficiency of operation of surface mining machines.

References

1. Kasztelewicz Z (2012) Bucket wheel excavators and spreaders. Technology of operation, Art-Tekst, Kraków
2. Pieczonka K (2007) Engineering of earth moving machines. Part 1. Basics of excavation, travel, hoisting and slewing. Wroclaw University of Technology Publishing House
3. Babiarz S, Dudek D (2007) Annals of failures and catastrophes of surface mining machines in Poland. Wroclaw University of Technology Publishing House
4. Bhattacharya J (ed) (2012) Design and selection of bulk material handling equipment and systems. Wide Publishing
5. Stamboliska Z, Rusiński E, Moczko P (2015) Proactive condition monitoring of low-speed machines. Springer International Publishing

Index

E. Rusiński et al., *Surface Mining Machines*,
DOI 10.1007/978-3-319-47792-3

MIX
Papier aus verantwortungsvollen Quellen
Paper from responsible sources
FSC® C105338

If you have any concerns about our products,
you can contact us on
ProductSafety@springernature.com

In case Publisher is established outside the EU,
the EU authorized representative is:
Springer Nature Customer Service Center GmbH
Europaplatz 3, 69115 Heidelberg, Germany

Printed by Libri Plureos GmbH
in Hamburg, Germany